A Course in SIMULATION

A Course in SIMULATION

Sheldon M. Ross
UNIVERSITY OF CALIFORNIA, BERKELEY

Macmillan Publishing Company
NEW YORK
Collier Macmillan Publishers
LONDON

Printed in the United States of America

Macmillan Publishing Company
866 Third Avenue, New York, New York 10022

Collier Macmillan Canada, Inc.

Library of Congress Cataloging-in-Publication Data

Ross, Sheldon M.
A course in simulation/Sheldon M. Ross.
p. cm.
ISBN 0-02-403891-1
1. Random variables. 2. Probabilities. 3. Computer simulation.
I. Title.
QA273.R82 1990 89-13546
519.2—dc20 CIP

Printing: 1 2 3 4 5 6 7 8 Year: 0 1 2 3 4 5 6 7 8 9

To Reb and Leesee

Preface

In formulating a stochastic model to describe a real phenomenon it used to be that one compromised between choosing a model that is a realistic replica of the actual situation and one whose mathematical analysis is tractable. That is, there did not seem to be any payoff in choosing a model that faithfully conformed to the phenomenon under study if it were not possible to mathematically analyze that model. Similar considerations have led to the concentration on asymptotic or steady-state results as opposed to the more useful ones on transient time. However, the relatively recent advent of fast and inexpensive computational power has opened up another approach—namely, to try to model the phenomenon as faithfully as possible and then to rely on a simulation study to analyze it.

In this text we show how to analyze a model by use of a simulation study. In particular we first show how a computer can be utilized to generate random (more precisely pseudo-random) numbers, and then how these random numbers can be used to generate the values of random variables from arbitrary distributions. Using the concept of discrete events we show how to use random variables to generate the behavior of a stochastic model over time. By continually generating the behavior of the system we show how to obtain estimators of desired quantities of interest. The statistical questions of when to stop a simulation and what confidence to place in the resulting estimators are considered. A variety of ways in which one can improve on the usual simulation estimators are presented. In addition, we show how to use simulation to determine whether the stochastic model chosen is consistent with a set of actual data.

The successive chapters in this text are as follows. Chapter 1 is an introductory chapter which presents a typical phenomenon that is of interest to study. Chapter 2 is a review of probability. Whereas this chapter is self-contained and does not assume the reader is familiar with probability, we imagine that it will indeed be a review for most readers. Chapter 3 deals with random numbers and how a variant of them (the so-called pseudo-random numbers) can be generated on a computer. The use of random numbers to generate discrete and then continuous random variables is considered in Chapters 4 and 5.

Chapter 6 presents the discrete event approach to track an arbitrary system as it evolves over time. Chapter 7 introduces the subject of statistics. Assuming our average reader has not previously studied statistics, the chapter starts with very basic concepts and ends by introducing the so-called bootstrap statistical approach which is quite useful in analyzing a simulation. Chapter 8 deals with the important subject of variance reduction. This is an attempt to improve on the usual simulation estimators by finding ones having the same mean and smaller variance. In successive sections of this chapter approaches to variance reduction as the use of antithetic variables, control variates, conditional expectations, statified sampling, importance sampling, and common random numbers are presented. Each of these approaches, sometimes in combination, is illustrated by a variety of examples. Chapter 9 is concerned with statistical validation techniques which are statistical procedures that can be used to validate the stochastic model when some real data are available. Goodness of fit tests such as the chi-square test and the Kolmogorov–Smirnov test are presented. Other sections in this chapter deal with the 2-sample and the n-sample problem and with ways of statistically testing the hypothesis that a given process is a Poisson process. Finally, in Chapter 10 we present some more advanced topics. Succeeding sections consider the alias method for generating discrete random variables, ways to generate two-dimensional Poisson processes, the Markov chain simulation method, and an important new approach to variance reduction which has never before appeared in a textbook and which makes use of the so-called random hazards.

There are exercises for the student at the end of each chapter. Also, there are some problems—written in BASIC so that they can be easily used on an IBM compatible personal computer—which are presented in an appendix of programs.

Sheldon M. Ross

ACKNOWLEDGMENTS

We thank Professors Karl Sigman, Columbia University, and Sadashiv Adiga, University of California, Berkeley, for their time and expertise in the review of the manuscript.

Contents

CHAPTER 1

Introduction

1

CHAPTER 2

Elements of Probability

5

CHAPTER 3

Random Numbers

36

CHAPTER 4

Generating Discrete Random Variables

44

CHAPTER 5

Generating Continuous Random Variables

59

CHAPTER 6

The Discrete Event Simulation Approach

81

C H A P T E R 7

Statistical Analysis of Simulated Data

93

C H A P T E R 8

Variance Reduction Techniques

112

C H A P T E R 9

Statistical Validation Techniques

148

C H A P T E R 10

Some Advanced Topics

Appendix of Programs

Index

A Course in SIMULATION

CHAPTER 1

Introduction

Consider the following situation faced by a pharmacist who is thinking of setting up a small pharmacy where he will fill prescriptions. He plans on opening his shop at 9 A.M. every weekday and expects that, on average, there will be about 32 prescriptions called in daily before 5 P.M. His experience indicates that the time it will take him to fill a prescription is a random quantity having a mean and standard deviation of 10 and 4 minutes, respectively. He plans on accepting no new prescriptions after 5 P.M., although he will remain in the shop past this time if necessary to fill all the prescriptions ordered that day. Given this scenario the pharmacist is probably, among other things interested in the answers to the following questions:

1. What is the average time that he will depart his store at night?
2. What proportion of days will he still be working at 5:30 P.M.?
3. What is the average time it will take him to fill a prescription (taking into account that he cannot necessarily immediately begin working on a newly arrived prescription but rather it must wait for him to complete work on all earlier ones)?
4. What proportion of prescriptions will be filled within 30 minutes?
5. If he changes his policy on accepting all prescriptions between 9 A.M. and 5 P.M. but rather only accepts new ones when there are fewer than five prescriptions still needing to be filled, how many prescriptions, on average, will be lost?
6. How would the conditions of limiting orders affect the answers to questions 1 through 4?

In order to employ mathematics to analyze this situation and answer the questions, we first construct a probability model. To do this it is necessary to make some reasonably accurate assumptions concerning the above scenario. For instance, we must make some assumptions about the probabilistic mechanism that describes the arrivals of the daily average of 32 customers. One possible assumption might be that the arrival rate is, in a probabilistic sense, constant over the day, whereas a second (probably more realistic) possible assumption is that the arrival rate depends on the time of day. We must then specify a probability distribution (having mean 10 and standard deviation 4) for the time it takes to service a prescription; and we must make assumptions about whether or not the service time of a given prescription always has this distribution or whether it changes as a function of other variables (e.g., the number of waiting prescriptions to be filled or the time of day). That is, we must make probabilistic assumptions about the daily arrival and service times. We must also decide if the probability law describing a given day changes as a function of the day of the week or whether it remains basically constant over time. After these assumptions, and possibly others, have been specified, a probability model of our scenario will have been constructed.

Once a probability model has been constructed, the answers to the questions can, in theory, be analytically determined. However, in practice, these questions are much too difficult to determine analytically and so to answer them we usually have to perform a simulation study. Such a study programs the probabilistic mechanism on a computer and by utilizing "random numbers" it simulates possible occurrences from this model over a large number of days and then utilizes the theory of statistics to estimate the answers to questions such as those given. In other words, the computer program utilizes random numbers to generate the values of random variables, having the assumed probability distributions, which represent the arrival times and the service times of prescriptions. Using these values, it then determines over many days the quantities of interest related to the questions. It then uses statistical techniques to provide estimated answers—for example, if out of 1000 simulated days, there are 122 in which the pharmacist is still working at 5:30 we would estimate that the answer to question 2 is .122.

In order to be able to execute such an analysis, one must have some knowledge of probability so as to decide on certain probability distributions and questions such as whether appropriate random variables are to be assumed independent or not. A review of probability is provided in Chapter 2. The bases of a simulation study are so-called random numbers. A discussion of these quantities and how they are computer generated is presented in Chapter 3. Chapters 4 and 5 show how one can use random numbers to generate the values of random variables having arbitrary distributions. Discrete distributions are considered in Chapter 4 and continuous ones in Chapter 5. After completing Chapter 5, the reader should have some insight into the construction of a probability model for a given system and also how to use random numbers to generate the values of random quantities related to this model. The use of these generated values to track the system as it evolves continuously over time—that is, the actual simulation of the system—is discussed in Chapter 6, where we present the concept of "discrete events" and indicate how to utilize these entities to obtain a systematic approach to simulating systems. The discrete event simulation approach leads to a computer program, which can be written in whatever language the reader is comfortable in, that simulates the system a large number of times. Some hints concerning the verification of this program—to ascertain that it is actually doing what is desired—are also given in Chapter 6. The use of the outputs of a simulation study to answer probabilistic questions concerning the model necessitates the use of the theory of statistics, and this subject is introduced in Chapter 7. This chapter starts with the simplest and most basic concepts in statistics and continues toward the recent innovation of "bootstrap statistics," which is quite useful in simulation. Our study of statistics indicates the importance of the variance of the estimators obtained from a simulation study as an indication of the efficiency of the simulation. In particular, the smaller this variance is, the smaller is the amount of simulation needed to obtain a fixed precision. As a result we are led, in Chapter 8, to ways of obtaining new estimators that are improvements over the raw simulation estimators because they have reduced variances. This topic of variance reduction is extremely important in a simulation study because it can substantially improve its efficiency. Chapter 9 shows how one can use the results of a simulation to verify, when some real life data are available, the appropriateness of the probability model (which we have simulated) to the real world situation. Finally, Chapter 10 considers some advanced techniques related both to the generation of random processes and to the use of this generation to estimate desired parameters.

Problems

1. The following data yield the arrival times and service times that each customer will require for the first 13 customers at a single server system. Upon arrival, a customer either enters service if the server is free or joins the waiting line. When the server completes work on a customer, the next one in line (i.e., the one who has been waiting the longest) enters service.

Arrival Times:	12	31	63	95	99	154	198	221	304	346	411	455	537
Service Times:	40	32	55	48	18	50	47	18	28	54	40	72	12

(a) Determine the departure times of these 13 customers.
(b) Repeat (a) when there are two servers and a customer can be served by either one.
(c) Repeat (a) under the new assumption that when the server completes a service, the next customer to enter service is the one that has been waiting the least time.

2. Consider a service station where customers arrive and are served in their order of arrival. Let A_n, S_n, and D_n denote, respectively, the arrival time, the service time, and the departure time of customer n. Suppose there is a single server and that the system is initially empty of customers.

(a) Argue that $D_0 = 0$, and for $n > 0$

$$D_n - S_n = \text{Maximum}\{A_n, D_{n-1}\}$$

(b) Determine the corresponding recursion formula when there are two servers.

(c) Determine the corresponding recursion formula when there are k servers.

(d) Write a computer program to determine the departure times as a function of the arrival and service times and use it to check your answers in parts (a) and (b) of Problem 1.

CHAPTER 2

Elements of Probability

2.1. Sample Space and Events

Consider an experiment whose outcome is not known in advance. Let S, called the sample space of the experiment, denote the set of all possible outcomes. For example, if the experiment consists of the running of a race among the seven horses numbered 1 through 7 then

$$S = \{\text{all orderings of } (1, 2, 3, 4, 5, 6, 7)\}$$

The outcome (3, 4, 1, 7, 6, 5, 2) means, for example, that the number 3 horse came in first, the number 4 horse came in second, and so on.

Any subset A of the sample space is known as an event. That is, an event is a set consisting of possible outcomes of the experiment. If the outcome of the experiment is contained in A, we say that A has occurred. For example, in the above if

$$A = \{\text{all outcomes in } S \text{ starting with } 5\}$$

then A is the event that the number 5 horse comes in first.

For any two events A and B we define the new event $A \cup B$, called the union of A and B, to consist of all outcomes that are either in A or B or in both A and B. Similarly, we define the event, AB, called the intersection of A and B, to consist of all outcomes that are in both A and B. That is, the event $A \cup B$ occurs if either A or B occurs; whereas the event AB occurs if both A and B occur. We can also define unions and intersections of more than two events. In particular, the union of the events $A_1, \ldots, A_n$—designated by $\cup_{i=1}^{n} A_i$—is defined to consist of all outcomes that are in at least one of the A_i. Similarly, the intersection of the events $A_1, \ldots, A_n$—designated by $A_1 A_2 \cdots A_n$—is defined to consist of all outcomes that are in all of the A_i.

For any event A we define the event A^c, referred to as the complement of A, to consist of all outcomes in the sample space S that are not in A. That is,

A^c occurs if and only if A does not. Since the outcome of the experiment must lie in the sample space S, it follows that S^c does not contain any outcomes and thus cannot occur. We call S^c the null set and designate it by $\varnothing$. If $AB = \varnothing$ so that A and B cannot both occur (since there are no outcomes that are in both A and B), we say that A and B are mutually exclusive.

2.2. Axioms of Probability

Suppose that for each event A of an experiment having sample space S there is a number, denoted by $P(A)$ and called the probability of the event A, which is in accord with the following three axioms:

Axiom 1:

$$0 \leq P(A) \leq 1$$

Axiom 2:

$$P(S) = 1$$

Axiom 3:

For any sequence of mutually exclusive events A_1, $A_2, \ldots$

$$P\left(\bigcup_{i=1}^{n} A_i\right) = \sum_{i=1}^{n} P(A_i), \qquad n = 1, 2, \ldots, \infty$$

Thus Axiom 1 states that the probability that the outcome of the experiment lies within A is some number between 0 and 1. Axiom 2 states that with probability 1 this outcome is a member of the sample space; and Axiom 3 states that for any set of mutually exclusive events, the probability that at least one of these events occurs is equal to the sum of their respective probabilities.

These three axioms can be used to prove a variety of results about probabilities. For instance, since A and A^c are always mutually exclusive, and since $A \cup A^c = S$, we have from Axioms 2 and 3 that

$$1 = P(S) = P(A \cup A^c) = P(A) + P(A^c)$$

or equivalently

$$P(A^c) = 1 - P(A)$$

In words, the probability that an event does not occur is 1 minus the probability that it does.

2.3. Conditional Probability and Independence

Consider an experiment that consists in flipping a coin twice, noting each time whether the result was heads or tails. The sample space of this experiment can be taken to be the following set of four outcomes:

$$S = \{(\text{H, H}), (\text{H, T}), (\text{T, H}), (\text{T, T})\}$$

where (H, T) means, for example, that the first flip lands heads and the second tails. Suppose now that each of the four possible outcomes is equally likely to occur and thus has probability $\frac{1}{4}$. Suppose further that we observe that the first flip lands on heads. Then, given this information, what is the probability that both flips land on heads? To calculate this probability we reason as follows: given that the initial flip lands heads, there can be at most two possible outcomes of our experiment, namely, (H, H) or (H, T). In addition, as each of these outcomes originally had the same probability of occurring, they should still have equal probabilities. That is, given that the first flip lands heads, the (conditional) probability of each of the outcomes (H, H) and (H, T) is $\frac{1}{2}$, whereas the (conditional) probability of the other two outcomes is 0. Hence the desired probability is $\frac{1}{2}$.

If we let A and B denote, respectively, the events that both flips land on heads and the event that the first flip lands on heads, then the probability obtained above is called the conditional probability of A given that B has occurred and is denoted by

$$P(A \mid B)$$

A general formula for $P(A \mid B)$ that is valid for all experiments and events A and B can be obtained in the same manner as given previously. Namely, if the event B occurs, then in order for A to occur it is necessary that the actual occurrence be a point in both A and B; that is, it must be in AB. Now since we know that B has occurred, it follows that B becomes our new sample space and hence the probability that the event AB occurs will equal the probability of AB relative to the probability of B. That is,

$$P(A \mid B) = \frac{P(AB)}{P(B)}$$

As indicated by the coin flip example, $P(A \mid B)$, the conditional probability of A, given that B occurred, is not generally equal to $P(A)$, the unconditional probability of A. In other words, knowing that B has occurred generally changes the probability that A occurs (what if they were mutually exclusive?). In the special case where $P(A|B)$ is equal to $P(A)$, we say that A and B are independent. Since $P(A|B) = P(AB)/P(B)$, we see that A is independent of B if

$$P(AB) = P(A)P(B)$$

Since this relation is symmetric in A and B, it follows that whenever A is independent of B, B is independent of A.

2.4. Random Variables

When an experiment is performed we are sometimes primarily concerned in the value of some numerical quantity determined by the result. These quantities of interest that are determined by the results of the experiment are known as random variables.

The cumulative distribution function, or more simply the distribution function, F, of the random variable X is defined for any real number x by

$$F(x) = P\{X \leq x\}$$

A random variable that can take either a finite or at most a countable number of possible values is said to be discrete. For a discrete random variable X we define its probability mass function $p(x)$ by

$$p(x) = P\{X = x\}$$

If X is a discrete random variable that takes on one of the possible values $x_1, x_2, \ldots$, then, since X must take on one of these values, we have

$$\sum_{i=1}^{\infty} p(x_i) = 1$$

Example 2a Suppose that X takes on one of the values 1, 2, or 3. If

$$p(1) = \tfrac{1}{4}, \qquad p(2) = \tfrac{1}{3}$$

then, since $p(1) + p(2) + p(3) = 1$, it follows that $p(3) = \frac{5}{12}$ ■

Whereas a discrete random variable assumes at most a countable set of possible values, we often have to consider random variables whose set of possible values is an interval. We say that the random variable X is a continuous random variable if there is a nonnegative function $f(x)$ defined for all real numbers x and having the property that for any set C of real numbers

$$P\{X \in C\} = \int_C f(x)\,dx \tag{2.1}$$

The function f is called the probability density function of the random variable X.

The relationship between the cumulative distribution $F(\cdot)$ and the probability density $f(\cdot)$ is expressed by

$$F(a) = P\{X \in (-\infty, a]\} = \int_{-\infty}^{a} f(x)\,dx$$

Differentiating both sides yields

$$\frac{d}{da}F(a) = f(a)$$

That is, the density is the derivative of the cumulative distribution function. A somewhat more intuitive interpretation of the density function may be obtained from Equation (2.1) as follows:

$$P\left\{a - \frac{\epsilon}{2} \le X \le a + \frac{\epsilon}{2}\right\} = \int_{a-\epsilon/2}^{a+\epsilon/2} f(x)\,dx \approx \epsilon f(a)$$

when ϵ is small. In other words, the probability that X will be contained in an interval of length ϵ around the point a is approximately $\epsilon f(a)$. From this, we see that $f(a)$ is a measure of how likely it is that the random variable will be near a.

In many experiments we are interested not only in probability distribution functions of individual random variables but also in the relationships between two or more of them. In order to specify the relationship between two random variables, we define the joint cumulative probability distribution function of X and Y by

$$F(x, y) = P\{X \le x, Y \le y\}$$

Thus $F(x, y)$ specifies the probability that X is less than or equal to x and simultaneously Y is less than or equal to y.

If X and Y are both discrete random variables, then we define the joint probability mass function of X and Y by

$$p(x, y) = P\{X = x, Y = y\}$$

Similarly, we say that X and Y are jointly continuous, with joint probability density function $f(x, y)$ if for any sets of real numbers C and D

$$P\{X \in C, Y \in D\} = \iint_{\substack{x \in C \\ y \in D}} f(x, y)\,dx\,dy$$

The random variables X and Y are said to be independent if for any two sets of real numbers C and D

$$P\{X \in C,\ Y \in D\} = P\{X \in C\}\ P\{Y \in D\}$$

That is X and Y are independent if for all sets C and D the events $A = \{X \in C\}$, $B = \{Y \in D\}$ are independent. Loosely speaking, X and Y are independent if knowing the value of one of them does not affect the probability distribution of the other. Random variables that are not independent are said to be dependent.

Using the three axioms of probability, we can show that the random variables X and Y will be independent if and only if

$$P\{X \leq x,\ Y \leq y\} = P\{X \leq x\}\ P\{Y \leq y\}, \qquad \text{for all } x,\ y$$

That is, X and Y will be independent if and only if

$$F(x,\ y) = F_X(x)\ F_Y(y), \qquad \text{for all } x,\ y$$

where F_X and F_Y are the distribution functions of X and Y.

2.5. Expectation

One of the most useful concepts in probability is that of the expectation of a random variable. If X is a discrete random variable that takes on one of the possible values $x_1, x_2, \ldots$, then the *expectation* or *expected value* of X, also called the mean of X and denoted by $E[X]$, is defined by

$$E[X] = \sum_i x_i P\{X = x_i\} \tag{2.2}$$

In words, the expected value of X is a weighted average of the possible values that X can take on, each value being weighted by the probability that X assumes it. For example, if the probability mass function of X is given by

$$p(0) = \tfrac{1}{2} = p(1)$$

then

$$E[X] = 0(\tfrac{1}{2}) + 1(\tfrac{1}{2}) = \tfrac{1}{2}$$

is just the ordinary average of the two possible values 0 and 1 that X can assume. On the other hand, if

$$p(0) = \tfrac{1}{3}, \qquad p(1) = \tfrac{2}{3}$$

then

$$E[X] = 0(\tfrac{1}{3}) + 1(\tfrac{2}{3}) = \tfrac{2}{3}$$

is a weighted average of the two possible values 0 and 1 where the value 1 is given twice as much weight as the value 0 since $p(1) = 2p(0)$.

Example 2b If I is an indicator random variable for the event A, that is, if

$$I = \begin{cases} 1 & \text{if } A \text{ occurs} \\ 0 & \text{if } A \text{ does not occur} \end{cases}$$

then

$$E[I] = 1P(A) + 0P(A^c) = P(A)$$

Hence the expectation of the indicator random variable for the event A is just the probability that A occurs. ■

If X is a continuous random variable having probability density function f, then, analogous to Equation (2.2), we define the expected value of X by

$$E[X] = \int_{-\infty}^{\infty} xf(x)\, dx$$

Example 2c If the probability density function of X is given by

$$f(x) = \begin{cases} 3x^2 & \text{if } 0 < x < 1 \\ 0 & \text{otherwise} \end{cases}$$

then

$$E[X] = \int_0^1 3x^3\, dx = \tfrac{3}{4}$$ ■

Suppose now that we wanted to determine the expected value not of the random variable X but of the random variable $g(X)$, where g is some given function. Since $g(X)$ takes on the value $g(x)$ when X takes on the value x, it seems intuitive that $E[g(X)]$ should be a weighted average of the possible values $g(x)$ with, for a given x, the weight given to $g(x)$ being equal to the probability (or probability density in the continuous case) that X will equal x. Indeed, the preceding can be shown to be true and we thus have the following result.

Proposition:

If X is a discrete random variable having probability mass function $p(x)$, then

$$E[g(X)] = \sum_x g(x)p(x)$$

whereas if X is continuous with probability density function $f(x)$, then

$$E[g(X)] = \int_{-\infty}^{\infty} g(x) f(x)\, dx$$

A consequence of the above proposition is the following.

Corollary:

If a and b are constants, then

$$E[aX + b] = aE[X] + b$$

Proof. In the discrete case

$$\begin{aligned} E[aX + b] &= \sum_x (ax + b)p(x) \\ &= a\sum_x xp(x) + b\sum_x p(x) \\ &= aE[X] + b \end{aligned}$$

Since the proof in the continuous case is similar, the result is established. □

It can be shown that expectation is a linear operation in the sense that for any two random variables X_1 and X_2

$$E[X_1 + X_2] = E[X_1] + E[X_2]$$

which easily generalizes to give

$$E\left[\sum_{i=1}^{n} X_i\right] = \sum_{i=1}^{n} E[X_i]$$

2.6. Variance

Whereas $E[X]$, the expected value of the random variable X, is a weighted average of the possible values of X, it yields no information about the variation of these values. One way of measuring this variation is to consider the average value of the square of the difference between X and $E[X]$. We are thus led to the following definition.

Definition:

If X is a random variable with mean μ, then the variance of X, denoted by $Var(X)$, is defined by

$$\text{Var}(X) = E[(X - \mu)^2]$$

An alternative formula for Var(X) is derived as follows:

$$\begin{aligned}\text{Var}(X) &= E[(X - \mu)^2] \\ &= E[X^2 - 2\mu X + \mu^2] \\ &= E[X^2] - E[2\mu X] + E[\mu^2] \\ &= E[X^2] - 2\mu E[X] + \mu^2 \\ &= E[X^2] - \mu^2\end{aligned}$$

That is,

$$\text{Var}(X) = E[X^2] - (E[X])^2$$

A useful identity, whose proof is left as an exercise, is that for any constants a and b

$$\text{Var}(aX + b) = a^2\text{Var}(X)$$

Whereas the expected value of a sum of random variables is equal to the sum of the expectations, the corresponding result is not, in general, true for variances. It is, however, true in the important special case where the random variables are independent. Before proving this let us define the concept of the covariance between two random variables.

Definition

The covariance of two random variables X and Y, denoted $Cov(X, Y)$, is defined by

$$\text{Cov}(X, Y) = E[(X - \mu_x)(Y - \mu_y)]$$

where $\mu_x = E[X]$ and $\mu_y = E[Y]$.

A useful expression for Cov(X, Y) is obtained by expanding the right side of the above equation and then making use of the linearity of expectation. This yields

$$\begin{aligned}\text{Cov}(X, Y) &= E[XY - \mu_x Y - X\mu_y + \mu_x\mu_y] \\ &= E[XY] - \mu_x E[Y] - E[X]\,\mu_y + \mu_x\mu_y \\ &= E[XY] - E[X]\,E[Y] \qquad (2.3)\end{aligned}$$

We now derive an expression for Var($X + Y$) in terms of their individual variances and the covariance between them. Since

$$E[X + Y] = E[X] + E[Y] = \mu_x + \mu_y$$

we see that

$$\begin{aligned}\text{Var}(X + Y) &= E[(X + Y - \mu_x - \mu_y)^2] \\ &= E[(X - \mu_x)^2 + (Y - \mu_y)^2 + 2(X - \mu_x)(Y - \mu_y)] \\ &= E[(X - \mu_x)^2] + E[(Y - \mu_y)^2] + 2E[(X - \mu_x)(Y - \mu_y)] \\ &= \text{Var}(X) + \text{Var}(Y) + 2\,\text{Cov}(X, Y) \qquad (2.4)\end{aligned}$$

We end this section by showing that the variance of the sum of independent random variables is equal to the sum of their variances.

Proposition:

If X and Y are independent random variables then

$$\text{Cov}(X, Y) = 0$$

and so, from Equation (2.4),

$$\text{Var}(X + Y) = \text{Var}(X) + \text{Var}(Y)$$

Proof. From Equation (2.3) it follows that we need to show that $E[XY] = E[X]E[Y]$. Now in the discrete case,

$$\begin{aligned}E[XY] &= \sum_j \sum_i x_i y_j P\{X = x_i,\ Y = y_j\} \\ &= \sum_j \sum_i x_i y_j P\{X = x_i\} P\{Y = y_j\} \qquad \text{by independence} \\ &= \sum_j y_j P\{Y = y_j\} \sum_i x_i P\{X = x_i\} \\ &= E[Y]E[X]\end{aligned}$$

Since a similar argument holds in the continuous case, the result is proved. □

2.7. Chebyshev's Inequality and the Laws of Large Numbers

We start by a result known as Markov's inequality.

Proposition: Markov's Inequality

If X takes on only nonnegative values, then for any value $a > 0$

$$P\{X \geq a\} \leq \frac{E[X]}{a}$$

Proof. We give a proof for the case where X is a continuous random variable having density function f.

$$\begin{aligned} E[X] &= \int_0^\infty xf(x)\,dx \\ &= \int_0^a xf(x)\,dx + \int_a^\infty xf(x)\,dx \\ &\geq \int_a^\infty xf(x)\,dx \\ &\geq \int_a^\infty af(x)\,dx \qquad \text{since } xf(x) \geq af(x) \text{ when } x \geq a \\ &= a\int_a^\infty f(x)\,dx = aP\{X \geq a\} \end{aligned}$$

and the result is proved. □

As a corollary we have the so-called Chebyshev's inequality, which states that the probability that a random variable differs from its mean by more than k of its standard deviations is bounded by $1/k^2$, where the standard deviation of a random variable is defined to be the square root of its variance.

Corollary: Chebyshev's Inequality

If X is a random variable having mean μ and variance σ^2 then for any value $k > 0$.

$$P\{|X - \mu| \geq k\sigma\} \leq \frac{1}{k^2}$$

Proof. Since $(X - \mu)^2/\sigma^2$ is a nonnegative random variable whose mean is

$$E\left[\frac{(X - \mu)^2}{\sigma^2}\right] = \frac{E[(X - \mu)^2]}{\sigma^2} = 1$$

we obtain from Markov's inequality that

$$P\left\{\frac{(X - \mu)^2}{\sigma^2} \geq k^2\right\} \leq \frac{1}{k^2}$$

The result now follows since the inequality $(X - \mu)^2/\sigma^2 \geq k^2$ is equivalent to the inequality $|X - \mu| \geq k\sigma$. □

We now use Chebyshev's inequality to prove the weak law of large numbers, which states that the probability that the average of the first n terms of a sequence of independent and identically random variables differs from its mean by more than ϵ goes to 0 as n goes to infinity.

Theorem: The Weak Law of Large Numbers

Let $X_1, X_2, \ldots$ be a sequence of independent and identically distributed random variables having mean μ. Then, for any $\epsilon > 0$,

$$P\left\{\left|\frac{X_1 + \cdots + X_n}{n} - \mu\right| > \epsilon\right\} \to 0 \qquad \text{as } n \to \infty$$

Proof. We give a proof under the additional assumption that the random variables X_i have a finite variance σ^2. Now

$$E\left[\frac{X_1 + \cdots + X_n}{n}\right] = \frac{1}{n}(E[X_1] + \cdots + E[X_n]) = \mu$$

and

$$\text{Var}\left(\frac{X_1 + \cdots + X_n}{n}\right) = \frac{1}{n^2}[\text{Var}(X_1) + \cdots + \text{Var}(X_n)] = \frac{\sigma^2}{n}$$

where the above equation makes use of the fact that the variance of the sum of independent random variables is equal to the sum of their variances. Hence, from Chebyshev's inequality, it follows that for any positive k

$$P\left\{\left|\frac{X_1 + \cdots + X_n}{n} - \mu\right| \geq \frac{k\sigma}{\sqrt{n}}\right\} \leq \frac{1}{k^2}$$

Hence, for any $\epsilon > 0$, by letting k be such that $k\sigma/\sqrt{n} = \epsilon$, that is, by letting $k^2 = n\epsilon^2/\sigma^2$, we see that

$$P\left\{\left|\frac{X_1 + \cdots + X_n}{n} - \mu\right| \geq \epsilon\right\} \leq \frac{\sigma^2}{n\epsilon^2}$$

which establishes the result. □

A generalization of the weak law is the strong law of large numbers, which states that, with probability 1,

$$\lim_{n\to\infty} \frac{X_1 + \cdots + X_n}{n} = \mu$$

That is, with certainty, the long run average of a sequence of independent and identically distributed random variables will converge to its mean.

2.8. Some Discrete Random Variables

There are certain types of random variables that frequently appear in applications. In this section we survey some of the discrete ones.

Binomial Random Variables

Suppose that n independent trials, each of which results in a "success" with probability p, are to be performed. If X represents the number of successes that occur in the n trials, then X is said to be a binomial random variable with parameters (n, p). Its probability mass function is given by

$$p_i \equiv P\{X = i\} = \binom{n}{i} p^i(1 - p)^{n-i}, \qquad i = 0, 1, \ldots, n \tag{2.5}$$

where $\binom{n}{i} = n!/[(n - i)!i!]$ is the binomial coefficient, equal to the number of different subsets of i elements that can be chosen from a set of n elements.

The validity of Equation (2.5) can be seen by first noting that the probability of any particular sequence of outcomes that results in i successes and $n - i$ failures is, by the assumed independence of trials, $p^i(1-p)^{n-i}$. Equation (2.5) then follows since there are $\binom{n}{i}$ different sequences of the n outcomes that result in i successes and $n - i$ failures—which can be seen by noting that there are $\binom{n}{i}$ different choices of the i trials that result in successes.

A binomial $(1, p)$ random variable is called a Bernoulli random variable.

Since a binomial (n, p) random variable X represents the number of successes in n independent trials, each of which results in a success with probability p, we can represent it as follows:

$$X = \sum_{i=1}^{n} X_i \tag{2.6}$$

where

$$X_i = \begin{cases} 1 & \text{if the } i\text{th trial is a success} \\ 0 & \text{otherwise} \end{cases}$$

Now

$$\begin{aligned} E[X_i] &= P\{X_i = 1\} = p \\ \operatorname{Var}(X_i) &= E[X_i^2] - E([X_i])^2 \\ &= p - p^2 = p(1-p) \end{aligned}$$

where the above equation uses the fact that $X_i^2 = X_i$ (since $0^2 = 0$ and $1^2 = 1$). Hence the representation (2.6) yields that, for a binomial (n, p) random variable X,

$$E[X] = \sum_{i=1}^{n} E[X_i] = np$$

$$\begin{aligned} \operatorname{Var}(X) &= \sum_{i=1}^{n} \operatorname{Var}(X_i) \qquad \text{since the } X_i \text{ are independent} \\ &= np(1-p) \end{aligned}$$

The following recursive formula expressing p_{i+1} in terms of p_i is useful when computing the binomial probabilities:

$$\begin{aligned} p_{i+1} &= \frac{n!}{(n-i-1)!\,(i+1)!} p^{i+1} (1-p)^{n-i-1} \\ &= \frac{n!\,(n-i)}{(n-i)!i!\,(i+1)} p^i (1-p)^{n-i} \frac{p}{1-p} \\ &= \frac{n-i}{i+1} \frac{p}{1-p} p_i \end{aligned}$$

Poisson Random Variables

A random variable X that takes on one of the values 0, 1, 2, . . . is said to be a Poisson random variable with parameter λ, $\lambda > 0$, if its probability mass function is given by

$$p_i = P\{X = i\} = e^{-\lambda}\frac{\lambda^i}{i!}, \qquad i = 0, 1, \ldots$$

The symbol e, defined by $e = \lim_{n\to\infty}(1 + 1/n)^n$, is a famous constant in mathematics that is roughly equal to 2.7183.

Poisson random variables have a wide range of applications. One reason for this is that such random variables may be used to approximate the distribution of the number of successes in a large number of trials (which are either independent or at most "weakly dependent") when each trial has a small probability of being a success. To see why this is so, suppose that X is a binomial random variable with parameters (n, p)—and so represents the number of successes in n independent trials when each trial is a success with probability p—and let $\lambda = np$. Then

$$\begin{aligned} P\{X = i\} &= \frac{n!}{(n-i)!\,i!}p^i(1-p)^{n-i} \\ &= \frac{n!}{(n-i)!\,i!}\left(\frac{\lambda}{n}\right)^i\left(1 - \frac{\lambda}{n}\right)^{n-i} \\ &= \frac{n(n-1)\cdots(n-i+1)}{n^i}\,\frac{\lambda^i}{i!}\,\frac{(1-\lambda/n)^n}{(1-\lambda/n)^i} \end{aligned}$$

Now for n large and p small,

$$\left(1 - \frac{\lambda}{n}\right)^n \approx e^{-\lambda}, \qquad \frac{n(n-1)\cdots(n-i+1)}{n^i} \approx 1, \qquad \left(1 - \frac{\lambda}{n}\right)^i \approx 1$$

Hence, for n large and p small,

$$P\{X = i\} \approx e^{-\lambda}\frac{\lambda^i}{i!}$$

Since the mean and variance of a binomial random variable Y are given by

$$E[Y] = np, \qquad \text{Var}(Y) = np(1-p) \approx np \qquad \text{for small } p$$

it is intuitive, given the relationship between binomial and Poisson random variables, that for a Poisson random variable X, having parameter λ,

$$E[X] = \text{Var}(X) = \lambda$$

An analytic proof of the above is left as an exercise.

To compute the Poisson probabilities we make use of the following recursive formula.

$$\frac{p_{i+1}}{p_i} = \frac{\dfrac{e^{-\lambda}\lambda^{i+1}}{(i+1)!}}{\dfrac{e^{-\lambda}\lambda^i}{i!}} = \frac{\lambda}{i+1}$$

or, equivalently,

$$p_{i+1} = \frac{\lambda}{i+1} p_i, \qquad i \geq 0$$

Geometric Random Variables

Consider independent trials, each of which is a success with probability p. If X represents the number of the first trial that is a success then

$$P\{X = n\} = p(1-p)^{n-1}, \qquad n \geq 1 \tag{2.7}$$

which is easily obtained by noting that in order for the first success to occur on the nth trial the first $n - 1$ must all be failures and the nth a success. Equation (2.7) now follows since the trials are independent.

A random variable whose probability mass function is given by (2.7) is said to be a geometric random variable with parameter p. The mean of the geometric is obtained as follows:

$$E[X] = \sum_{n=1}^{\infty} np(1-p)^{n-1} = \frac{1}{p}$$

where the above equation made use of the algebraic identity, for $0 < x < 1$,

$$\sum_{n=1}^{\infty} nx^{n-1} = \frac{d}{dx}\left(\sum_{n=0}^{\infty} x^n\right) = \frac{d}{dx}\left(\frac{1}{1-x}\right) = \frac{1}{(1-x)^2}$$

It is also not difficult to show that

$$\mathrm{Var}(X) = \frac{1-p}{p^2}$$

The Negative Binomial Random Variable

If we let X denote the number of trials needed to amass a total of r successes when each trial is independently a success with probability p, then X is said to be a negative binomial, sometimes called a Pascal, random variable with parameters p and r. The probability mass function of such a random variable is given by the following:

$$P\{X = n\} = \binom{n-1}{r-1} p^r(1-p)^{n-r}, \qquad n \geq r \tag{2.8}$$

To see why Equation (2.8) is valid note that in order for it to take exactly n trials to amass r successes, the first $n - 1$ trials must result in exactly $r - 1$ successes—and the probability of this is $\binom{n-1}{r-1} p^{r-1}(1-p)^{n-r}$—and then the nth trial must be a success—and the probability of this is p.

If we let X_i, $i = 1, \ldots, r$, denote the number of trials needed after the $(i - 1)$st success to obtain the ith success, then it is easy to see that they are independent geometric random variables with common parameter p. Since

$$X = \sum_{i=1}^{r} X_i$$

we see that

$$E[X] = \sum_{i=1}^{r} E[X_i] = \frac{r}{p}$$

$$\text{Var}(X) = \sum_{i=1}^{r} \text{Var}(X_i) = \frac{r(1-p)}{p^2}$$

where the above made use of the corresponding result for geometric random variables.

Hypergeometric Random Variables

Consider an urn containing $N + M$ balls, of which N are light colored and M are dark colored. If a sample of size n is randomly chosen [in the sense that each of the $\binom{N+M}{n}$ subsets of size n is equally likely to be chosen] then X, the number of light colored balls selected, has probability mass function

$$P\{X = i\} = \frac{\binom{N}{i}\binom{M}{n-i}}{\binom{N+M}{n}}$$

A random variable X whose probability mass function is given by the above equation is called a hypergeometric random variable.

Suppose that the n balls are chosen sequentially. If we let

$$X_i = \begin{cases} 1 & \text{if the } i\text{th selection is light} \\ 0 & \text{otherwise} \end{cases}$$

then

$$X = \sum_{i=1}^{n} X_i \tag{2.9}$$

and so

$$E[X] = \sum_{i=1}^{n} E[X_i] = \frac{nN}{N + M}$$

where the above equation uses the fact that, by symmetry, the ith selection is equally likely to be any of the $N + M$ balls, and so $E[X_i] = P\{X_i = 1\} = N/(N + M)$.

Since the X_i are not independent (why not?), the utilization of the representation (2.9) to compute Var(X) involves covariance terms. The end product can be shown to yield the result

$$\text{Var}(X) = \frac{nNM}{(N + M)^2}\left(1 - \frac{n - 1}{N + M - 1}\right)$$

2.9. Continuous Random Variables

In this section we consider certain types of continuous random variable.

Uniformly Distributed Random Variables

A random variable X is said to be uniformly distributed over the interval (a, b), $a < b$, if its probability density function is given by

$$f(x) = \begin{cases} \dfrac{1}{(b - a)} & \text{if } a < x < b \\ 0 & \text{otherwise} \end{cases}$$

In other words, X is uniformly distributed over (a, b) if it puts all its mass on that interval and it is equally likely to be "near" any point on that interval.

The mean and variance of a uniform (a, b) random variable are obtained as follows:

$$E[X] = \frac{1}{b - a}\int_a^b x\,dx = \frac{b^2 - a^2}{2(b - a)} = \frac{b + a}{2}$$

$$E[X^2] = \frac{1}{b - a}\int_a^b x^2\,dx = \frac{b^3 - a^3}{3(b - a)} = \frac{a^2 + b^2 + ab}{3}$$

and so

$$\text{Var}(X) = \tfrac{1}{3}(a^2 + b^2 + ab) - \tfrac{1}{4}(a^2 + b^2 + 2ab) = \tfrac{1}{12}(b - a)^2$$

Thus for instance, the expected value is, as one might have expected, the midpoint of the interval (a, b).

The distribution function of X is given, for $a < x < b$, by

$$F(x) = P\{X \leq x\} = \int_a^x (b-a)^{-1}\, dx = \frac{x-a}{b-a}$$

Normal Random Variables

A random variable X is said to be normally distributed with mean μ and variance σ^2 if its probability density function is given by

$$f(x) = \frac{1}{\sqrt{2\pi}\,\sigma} e^{-(x-\mu)^2/2\sigma^2}, \qquad -\infty < x < \infty$$

The normal density is a bell-shaped curve that is symmetric about μ (see Figure 2.1).

It is not difficult to show that the parameters μ and σ^2 do indeed equal the expectation and variance of the normal. That is,

$$E[X] = \mu \qquad \text{and} \qquad \text{Var}(X) = \sigma^2$$

An important fact about normal random variables is that if X is normal with mean μ and variance σ^2, then for any constants a and b, $aX + b$ is normally distributed with mean $a\mu + b$ and variance $a^2\sigma^2$. It follows from this that if X is normal with mean μ and variance σ^2 then

$$Z = \frac{X-\mu}{\sigma}$$

is normal with mean 0 and variance 1. Such a random variable Z is said to have a unit normal distribution. Let ϕ denote the distribution function of a unit normal random variable; that is,

$$\phi(x) = \frac{1}{\sqrt{2\pi}} \int_{-\infty}^{x} e^{-x^2/2}\, dx, \qquad -\infty < x < \infty$$

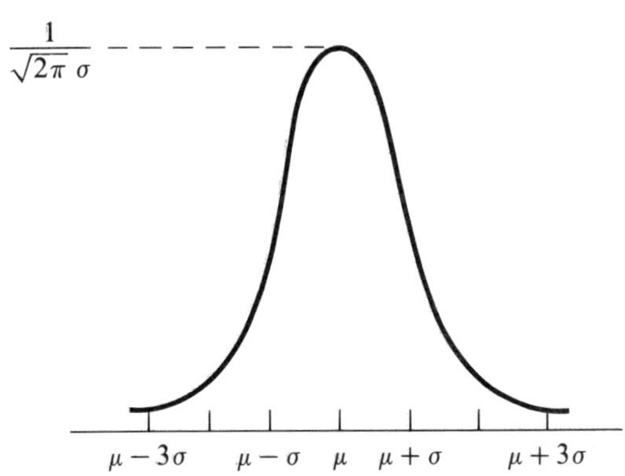

Figure 2.1 The normal density function.

The result that $Z = (X - \mu)/\sigma$ has a unit normal distribution when X is normal with mean μ and variance σ^2 is quite useful because it allows us to evaluate all probabilities concerning X in terms of ϕ. For example, the distribution function of X can be expressed as

$$\begin{aligned} F(x) &= P\{X \le x\} \\ &= P\left\{\frac{X - \mu}{\sigma} \le \frac{x - \mu}{\sigma}\right\} \\ &= P\left\{Z \le \frac{x - \mu}{\sigma}\right\} \\ &= \phi\left(\frac{x - \mu}{\sigma}\right) \end{aligned}$$

The value of $\phi(x)$ can be determined either by looking it up in a table or by writing a computer program to approximate it.

For a in the interval (0, 1), let z_a be such that

$$P\{Z > z_a\} = 1 - \phi(z_a) = a$$

That is, a unit normal will exceed z_a with probability a. (see Figure 2.2) The values of z_a can be obtained from a table of the values of ϕ. For example, since

$$\phi(1.64) = .95, \qquad \phi(1.96) = .975, \qquad \phi(2.33) = .99$$

we see that

$$z_{.05} = 1.64, \qquad z_{.025} = 1.96, \qquad z_{.01} = 2.33$$

The wide applicability of normal random variables results from one of the most important theorems of probability theory—the central limit theorem, which asserts that the sum of a large number of independent random variables has approximately a normal distribution. The simplest form of this remarkable theorem is as follows.

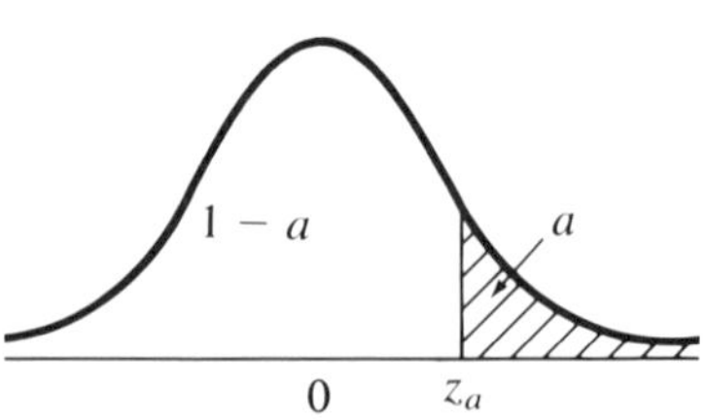

Figure 2.2 $P\{Z > z_a\} = a$.

The Central Limit Theorem

Let $X_1, X_2, \ldots$ be a sequence of independently and identically distributed random variables having finite mean μ and finite variance σ^2. Then

$$\lim_{n\to\infty} P\left\{\frac{X_1 + \cdots + X_n - n\mu}{\sigma\sqrt{n}} < x\right\} = \phi(x)$$

Exponential Random Variables

A continuous random variable having probability density function

$$f(x) = \lambda e^{-\lambda x}, \qquad 0 < x < \infty$$

for some $\lambda > 0$ is said to be an exponential random variable with parameter λ. Its cumulative distribution is given by

$$F(x) = \int_0^x \lambda e^{-\lambda x}\, dx = 1 - e^{-\lambda x}, \qquad 0 < x < \infty$$

It is easy to verify that the expected value and variance of such a random variable are as follows:

$$E[X] = \frac{1}{\lambda} \qquad \text{and} \qquad \operatorname{Var}(X) = \frac{1}{\lambda^2}$$

The key property of exponential random variables is that they possess the "memoryless property," where we say that the nonnegative random variable is memoryless if

$$P\{X > s + t \mid X > s\} = P\{X > t\} \qquad \text{for all } s, t \geq 0 \tag{2.10}$$

To understand why the above is called the memoryless property, imagine that X represents the lifetime of some unit, and consider the probability that a unit of age s will survive an additional time t. Since this will occur if the lifetime of the unit exceeds $t + s$ given that it is still alive at time s, we see that

$$P\{\text{additional life of an item of age } s \text{ exceeds } t\} = P\{X > s + t \mid X > s\}$$

Thus Equation (2.10) is a statement of fact that the distribution of the remaining life of an item of age s does not depend on s. That is, it is not necessary to remember the age of the unit to know its distribution of remaining life.

Equation (2.10) is equivalent to

$$P\{X > s + t\} = P\{X > s\}\, P\{X > t\}$$

As the above equation is satisfied whenever X is an exponential random variable—since, in this case, $P\{X > x\} = e^{-\lambda x}$—we see that exponential random variables are memoryless (and indeed it is not difficult to show that they are the only memoryless random variables).

Another useful property of exponential random variables is that they remain exponential when multiplied by a positive constant. To see this suppose that X is exponential with parameter λ, and let c be a positive number. Then

$$P\{cX \leq x\} = P\left\{X \leq \frac{x}{c}\right\} = 1 - e^{-\lambda x/c}$$

which shows that cX is exponential with parameter λ/c.

The Poisson Process and Gamma Random Variables

Suppose that "events" are occurring at random time points and let $N(t)$ denote the number of events that occur in the time interval $[0, t]$. These events are said to constitute a *Poisson process having rate* λ, $\lambda > 0$, if

(a) $N(0) = 0$
(b) The number of events that occur in disjoint time intervals are independent.
(c) The distribution of the number of events that occur in a given interval depends only on the length of the interval and not on its location.
(d) $\lim_{h\to 0} \dfrac{P\{N(h) = 1\}}{h} = \lambda$
(e) $\lim_{h\to 0} \dfrac{P\{N(h) \geq 2\}}{h} = 0$

Thus Condition (a) states that the process begins at time 0. Condition (b), the *independent increment* assumption, states that the number of events by time t [i.e., $N(t)$] is independent of the number of events that occur between t and $t + s$ [i.e., $N(t + s) - N(t)$]. Condition (c), the *stationary increment* assumption, states that the probability distribution of $N(t + s) - N(t)$ is the same for all values of t. Conditions (d) and (e) state that in a small interval of length h, the probability of one event occurring is approximately λh, whereas the probability of two or more is approximately 0.

We now argue that these assumptions imply that the number of events occurring in an interval of length t is a Poisson random variable with mean λt. To do so, consider the interval $[0, t]$, and break it up into n nonoverlapping subintervals of length t/n (Figure 2.3). Consider first the number of these subintervals that contain an event. As each subinterval independently [by Condition (b)] contains an event with the same probability [by Condition (c)], which is approximately equal to $\lambda t/n$, it follows that the number of such

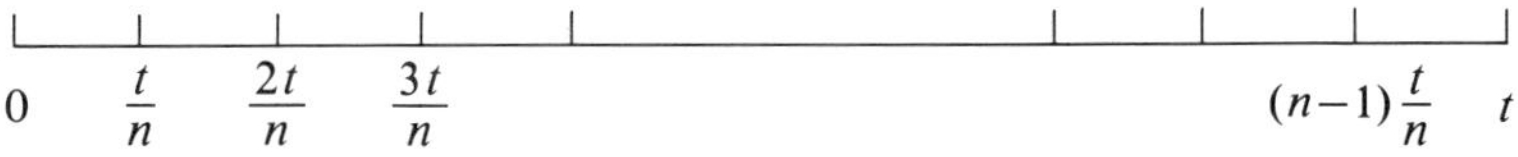

Figure 2.3

intervals is a binomial random variable with parameters n and $p \approx \lambda t/n$. Hence, by the argument yielding the convergence of the binomial to the Poisson, we see by letting $n \to \infty$ that the number of such subintervals converges to a Poisson random variable with mean λt. As it can be shown that Condition (e) implies that the probability that any of these subintervals contains two or more events goes to 0 as $n \to \infty$, it follows that $N(t)$, the number of events that occur in $[0, t]$, is a Poisson random variable with mean λt.

For a Poisson process let us denote by X_1 the time of the first event. Furthermore, for $n > 1$, let X_n denote the elapsed time between the $(n-1)$st and the nth event. The sequence $\{X_n, n = 1, 2, \ldots\}$ is called the *sequence of interarrival times*. For instance, if $X_1 = 5$ and $X_2 = 10$, then the first event of the Poisson process will occur at time 5 and the second at time 15.

We now determine the distribution of the X_n. To do so, we first note that the event $\{X_1 > t\}$ takes place if and only if no events of the Poisson process occur in the interval $[0, t]$, and thus

$$P\{X_1 > t\} = P\{N(t) = 0\} = e^{-\lambda t}$$

Hence X_1 has an exponential distribution with mean $1/\lambda$. To obtain the distribution of X_2, note that

$$\begin{aligned} P\{X_2 > t \mid X_1 = s\} &= P\{0 \text{ events in } (s, \; s + t] \mid X_1 = s\} \\ &= P\{0 \text{ events in } (s, \; s + t]\} \\ &= e^{-\lambda t} \end{aligned}$$

where the last two equations followed from independent and stationary increments. Therefore, from the foregoing, we conclude that X_2 is also an exponential random variable with mean $1/\lambda$ and, furthermore, that X_2 is independent of X_1. Repeating the same argument yields:

Proposition

The interarrival times $X_1, X_2, \ldots$ are independent and identically distributed exponential random variables with parameter λ.

Let $S_n = \sum_{i=1}^{n} X_i$ denote the time of the nth event. Since S_n will be less than or equal to t if and only if there have been at least n events by time t, we see that

$$P\{S_n \le t\} = P\{N(t) \ge n\}$$
$$= \sum_{j=n}^{\infty} e^{-\lambda t} \frac{(\lambda t)^j}{j!}$$

Since the left-hand side is the cumulative distribution function of S_n, we obtain, upon differentiation, that the density function of S_n—call it $f_n(t)$—is given by

$$f_n(t) = \sum_{j=n}^{\infty} \lambda e^{-\lambda t} \frac{j(\lambda t)^{j-1}}{j!} - \sum_{j=n}^{\infty} \lambda e^{-\lambda t} \frac{(\lambda t)^j}{j!}$$
$$= \sum_{j=n}^{\infty} \lambda e^{-\lambda t} \frac{(\lambda t)^{j-1}}{(j-1)!} - \sum_{j=n}^{\infty} \lambda e^{-\lambda t} \frac{(\lambda t)^j}{j!}$$
$$= \lambda e^{-\lambda t} \frac{(\lambda t)^{n-1}}{(n-1)!}$$

Definition:

A random variable having probability density function

$$f(t) = \lambda e^{-\lambda t} \frac{(\lambda t)^{n-1}}{(n-1)!}, \qquad t > 0$$

is said to be a gamma random variable with parameters (n, λ).

Thus we see that S_n, the time of the nth event of a Poisson process having rate λ, is a gamma random variable with parameters (n, λ). In addition, we obtain from the representation $S_n = \sum_{i=1}^{n} X_i$ and the previous proposition, which stated that these X_i are independent exponentials with rate λ, the following corollary.

Corollary:

The sum of n independent exponential random variables, each having parameter λ, *is a gamma random variable with parameters* (n, λ).

The Nonhomogeneous Poisson Process

From a modeling point of view the major weakness of the Poisson process is its assumption that events are just as likely to occur in all intervals of equal size. A generalization, which relaxes this assumption, leads to the so-called nonhomogeneous or nonstationary process.

If "events" are occurring randomly in time, and $N(t)$ denotes the number of events that occur by time t, then we say that $\{N(t),\ t \geq 0\}$ constitutes a nonhomogeneous Poisson process with intensity function $\lambda(t)$, $t \geq 0$, if

(a) $N(0) = 0$.
(b) The number of events that occur in disjoint time intervals are independent.
(c) $\lim_{h \to 0} P\{\text{exactly 1 event between } t \text{ and } t + h\}/h = \lambda(t)$.
(d) $\lim_{h \to 0} P\{2 \text{ or more events between } t \text{ and } t + h\}/h = 0$.

The function $m(t)$, defined by

$$m(t) = \int_0^t \lambda(s)\, ds, \qquad t \geq 0$$

is called the mean-value function. The following result can be established.

Proposition:

$N(t + s) - N(t)$ is a Poisson random variable with mean $m(t + s) - m(t)$.

The quantity $\lambda(t)$, called the intensity at time t, indicates how likely it is that an event will occur around the time t. [Note that when $\lambda(t) \equiv \lambda$ the nonhomogeneous reverts to the usual Poisson process.] The following proposition gives a useful way of interpreting a nonhomogeneous Poisson process.

Proposition:

Suppose that events are occurring according to a Poisson process having rate λ, and suppose that, independently of anything that came before, an event that occurs at time t is counted with probability $p(t)$. Then the process of counted events constitutes a nonhomogeneous Poisson process with intensity function $\lambda(t) = \lambda p(t)$.

Proof. This proposition is proved by noting that the previously given conditions are all satisfied. Conditions (a), (b), and (d) follow since the corresponding result is true for all (not just the counted) events. Condition (c) follows since

$$\begin{aligned} &P\{1 \text{ counted event between } t \text{ and } t+h\} \\ &\qquad = P\{1 \text{ event and it is counted}\} \\ &\qquad\quad + P\{2 \text{ or more events and exactly 1 is counted}\} \\ &\qquad \approx \lambda h p(t) \end{aligned}$$ □

2.10. Conditional Expectation and Conditional Variance

If X and Y are jointly discrete random variables, we define $E[X|Y = y]$, the conditional expectation of X given that $Y = y$, by

$$E[X|Y = y] = \sum_x xP\{X = x|Y = y\} = \frac{\sum_x xP\{X = x, Y = y\}}{P\{Y = y\}}$$

In other words, the conditional expectation of X, given that $Y = y$, is defined as before as a weighted average of all the possible values of X but now with the weight given to the value x being equal to the conditional probability that X equals x given that Y equals y.

Similarly, if X and Y are jointly continuous with joint density function $f(x, y)$, we define the conditional expectation of X, given that $Y = y$, by

$$E[X|Y = y] = \frac{\int xf(x, y)\, dx}{\int f(x, y)\, dx}$$

Let $E[X|Y]$ denote that function of the random variable Y whose value at $Y = y$ is $E[X|Y = y]$; and note that $E[X|Y]$ is itself a random variable. The following proposition is quite useful.

Proposition:

$$E[E[X|Y]] = E[X] \tag{2.11}$$

If y is a discrete random variable then Equation (2.11) states that

$$E[X] = \sum_y E[X|Y = y]P\{Y = y\}$$

whereas if Y is continuous with density g, then (2.11) states

$$E[X] = \int E[X|Y = y]g(y)\, dy$$

We now give a proof of the above proposition when X and Y are discrete:

$$\begin{aligned}
\sum_y E[X|Y = y]P\{Y = y\} &= \sum_y \sum_x xP\{X = x|Y = y\}P\{Y = y\} \\
&= \sum_y \sum_x xP\{X = x, Y = y\} \\
&= \sum_x x \sum_y P\{X = x, Y = y\} \\
&= \sum_x xP\{X = x\} \\
&= E[X]
\end{aligned}$$

We can also define the conditional variance of X, given the value of Y, as follows:

$$\text{Var}(X|Y) = E[(X - E[X|Y])^2|Y]$$

That is, $\text{Var}(X|Y)$ is a function of Y, which at $Y = y$ is equal to the variance of X given that $Y = y$. By the same reasoning that yields the identity $\text{Var}(X) = \text{E}[\text{X}^2] - (\text{E}[\text{X}])^2$ we have that

$$\text{Var}(X|Y) = E[X^2|Y] - (E[X|Y])^2$$

Taking expectations of both sides of the above equation gives

$$\begin{aligned}
E[\text{Var}(X|Y)] &= E[E[X^2|Y]] - E[(E[X|Y])^2] \\
&= E[X^2] - E[(E[X|Y])^2] \qquad (2.12)
\end{aligned}$$

Also, because $E[E[X|Y]] = E[X]$, we have that

$$\text{Var}(E[X|Y]) = E[(E[X|Y])^2] - (E[X])^2 \qquad (2.13)$$

Upon adding Equations (2.12) and (2.13) we obtain the following identity, known as the conditional variance formula.

THE CONDITIONAL VARIANCE FORMULA

$$\operatorname{Var}(X) = E[\operatorname{Var}(X|Y)] + \operatorname{Var}(E[X|Y])$$

Problems

1. Consider an experiment that consists of six horses, numbered 1 through 6, running a race, and suppose that the sample space is given by

$$S = \{\text{all orderings of } (1, 2, 3, 4, 5, 6)\}$$

Let A denote the event that the number 1 horse is among the top three finishers, let B denote the event that the number 2 horse comes in second, and let C denote the event that the number 3 horse comes in third.
(a) Describe the event $A \cup B$. How many outcomes are contained in this event?
(b) How many outcomes are contained in the event AB?
(c) How many outcomes are contained in the event ABC?
(d) How many outcomes are contained in the event $A \cup BC$?

2. (a) For any events A and B show that

$$A \cup B = A \cup A^cB$$
$$A = AB \cup AB^c$$

(b) Show that

$$P(A \cup B) = P(A) + P(B) - P(AB)$$

3. A couple has two children. What is the probability that both are girls given that the eldest is a girl? Assume that all four possibilities are equally likely.

4. The king comes from a family of two children. What is the probability that the other child is his brother?

5. The random variable X takes on one of the values 1, 2, 3, 4 with probabilities

$$P\{X = i\} = ic, \qquad i = 1, 2, 3, 4$$

for some value c. Find $P\{2 \leq X \leq 3\}$.

6. The continuous random variable X has a probability density function given by

$$f(x) = cx, \qquad 0 < x < 1$$

Find $P\{X > \frac{1}{2}\}$.

7. If X and Y have a joint probability density function specified by

$$f(x, y) = 2e^{-(x+2y)}, \qquad 0 < x < \infty,\ 0 < y < \infty$$

Find $P\{X < Y\}$.

8. Find the expected value of the random variable specified in Problem 5.

9. Find $E[X]$ for the random variable of Problem 6.

10. There are 10 different types of coupon and each time one obtains a coupon it is equally likely to be any of the 10 types. Let X denote the number of distinct types contained in a collection of N coupons, and find $E[X]$. [Hint: For $i = 1, \ldots, 10$ let

$$X_i = \begin{cases} 1 & \text{if a type } i \text{ coupon is among the } N \\ 0 & \text{otherwise} \end{cases}$$

and make use of the representation $X = \sum_{i=1}^{10} X_i$.

11. A die having six sides is rolled. If each of the six possible outcomes is equally likely, determine the variance of the number that appears.

12. Suppose that X has probability density function

$$f(x) = ce^x, \qquad 0 < x < 1$$

Determine $\text{Var}(X)$.

13. Show that $\text{Var}(aX + b) = a^2\text{Var}(X)$.

14. Suppose that X, the amount of liquid apple contained in a container of commercial apple juice, is a random variable having mean 4 grams.
(a) What can be said about the probability that a given container contains more than 6 grams of liquid apple?
(b) If $\text{Var}(X) = 4(\text{grams})^2$, what can be said about the probability that a given container will contain between 3 and 5 grams of liquid apple?

15. An airplane needs at least half of its engines to safely complete its mission. If each engine independently functions with probability p, for what values of p is a three-engine plane safer than a five-engine plane?

16. For a binomial random variable X with parameters (n, p), show that $P\{X = i\}$ first increases and then decreases, reaching its maximum value when i is the largest integer less than or equal to $(n + 1)p$.

17. If X and Y are independent binomial random variables with respective parameters (n, p) and (m, p), argue, without any calculations, that $X + Y$ is binomial with parameters $(n + m, p)$.

18. Explain why the following random variables all have approximately a Poisson distribution:
(a) The number of misprints in a given chapter of this book.

(b) The number of wrong telephone numbers dialed daily.
(c) The number of customers that enter a given post office on a given day.

19. If X is a Poisson random variable with parameter λ, show that
(a) $E[X] = \lambda$.
(b) $\mathrm{Var}(X) = \lambda$.

20. Let X and Y be independent Poisson random variables with respective parameters λ_1 and λ_2. Use the result of Problem 17 to heuristically argue that $X + Y$ is Poisson with parameter $\lambda_1 + \lambda_2$. Then give an analytic proof of this. [Hint:

$$P\{X + Y = k\} = \sum_{i=0}^{k} P\{X = i, Y = k - i\} = \sum_{i=0}^{k} P\{X = i\}P\{Y = k - i\}]$$

21. Explain how to make use of the relationship

$$p(i + 1) = \frac{\lambda}{i + 1} p(i)$$

to compute efficiently the Poisson probabilities.

22. Find $P\{X > n\}$ when X is a geometric random variable with parameter p.

23. Two players play a certain game until one has won a total of five games. If player A wins each individual game with probability .6, what is the probability she will win the match?

24. Consider the hypergeometric model of Section 2.8, and suppose that the white balls are all numbered. For $i = 1, \ldots, N$ let

$$Y_i = \begin{cases} 1 & \text{if white ball numbered } i \text{ is selected} \\ 0 & \text{otherwise} \end{cases}$$

Argue that $X = \Sigma_{i=1}^{N} Y_i$ and then use this representation to determine $E[X]$. Verify that it checks with the result given in Section 2.8.

25. The bus will arrive at a time that is uniformly distributed between 8 and 8:30 A.M. If we arrive at 8 A.M., what is the probability that we will wait between 5 and 15 minutes?

26. For a normal random variable with parameters μ and σ^2 show that
(a) $E[X] = \mu$.
(b) $\mathrm{Var}(X) = \sigma^2$.

27. Let X be a binomial random variable with parameters (n, p). Explain why

$$P\left\{\frac{X-np}{\sqrt{np(1-p)}} \le x\right\} \approx \frac{1}{\sqrt{2\pi}} \int_{-\infty}^{x} e^{-x^2/2}\,dx$$

when n is large.

28. If X is an exponential random variable with parameter λ, show that
(a) $E[X] = 1/\lambda$.
(b) $\mathrm{Var}(X) = 1/\lambda^2$.

29. Persons A, B, and C are waiting at a bank having two tellers when it opens in the morning. Persons A and B each go to a teller and C waits on line. If the time it takes to serve a customer is an exponential random variable with parameter λ, what is the probability that C is the last to leave the bank? [Hint: No computations are necessary.]

30. Consider a Poisson process in which events occur at a rate 0.3 per hour. What is the probability that no events occur between 10 A.M. and 2 P.M.?

31. For a Poisson process with rate λ, find $P\{N(s) = k \mid N(t) = n\}$ when $s < t$.

32. Repeat Problem 31 for $s > t$.

33. If X is a gamma random variable with parameters (n, λ), find
(a) $E[X]$.
(b) $\mathrm{Var}(X)$.

34. An urn contains four white and six black balls. A random sample of size 4 is chosen. Let X denote the number of white balls in the sample. An additional ball is now selected from the remaining six balls in the urn. Let Y equal 1 if this ball is white and 0 if it is black. Find
(a) $E[Y \mid X = 2]$.
(b) $E[X \mid Y = 1]$.
(c) $\mathrm{Var}(Y \mid X = 0)$.
(d) $\mathrm{Var}(X \mid Y = 1)$.

35. If X and Y are independent and identically distributed exponential random variables, show that the conditional distribution of X, given that $X + Y = t$, is the uniform distribution on $(0, t)$.

CHAPTER 3

Random Numbers

Introduction

The building block of a simulation study is the ability to generate random numbers, where a random number represents the value of a random variable uniformly distributed on (0, 1). In this chapter we explain how such numbers are computer generated and also begin to illustrate their uses.

3.1. Pseudo-Random Number Generation

Whereas random numbers were originally either manually or mechanically generated, by using such techniques as spinning wheels, or dice rolling, or card shuffling, the modern approach is to use a computer to successively generate pseudo-random numbers. These pseudo-random numbers constitute a sequence of values, which, although they are deterministically generated, have all the appearances of being independent uniform (0, 1) random variables.

One of the most common approaches to generating pseudo-random numbers starts with an initial value x_0, called the seed, and then recursively computes successive values x_n, $n \geq 1$, by letting

$$x_n = ax_{n-1} \quad \text{modulo } m \tag{3.1}$$

where a and m are given positive integers, and where the above means that ax_{n-1} is divided by m and the remainder is taken as the value of x_n. Thus each x_n is either $0, 1, \ldots, m-1$ and the quantity x_n/m—called a pseudo-random number—is taken as an approximation to the value of a uniform (0, 1) random variable.

The approach specified by Equation (3.1) to generate random numbers is called the multiplicative congruential method. Since each of the numbers x_n

assumes one of the values $0, 1, \ldots, m - 1$, it follows that after some finite number (of at most m) of generated values a value must repeat itself; and once this happens the whole sequence will begin to repeat. Thus we want to choose the constants a and m so that, for any initial seed x_0, the number of variables that can be generated before this repeating occurs is large.

In general the constants a and m should be chosen to satisfy three criteria:

1. For any initial seed, the resultant sequence has the "appearance" of being a sequence of independent uniform (0, 1) random variables.
2. For any initial seed, the number of variables that can be generated before repetition begins is large.
3. The values can be computed efficiently on a digital computer.

A guideline that appears to be of help in satisfying the above three conditions is that m should be chosen to be a large prime number that can be fitted to the computer word size. For a 32-bit word machine (where the first bit is a sign bit) it has been shown that the choices of $m = 2^{31} - 1$ and $a = 7^5 =$ 16,807 result in desirable properties. (For a 36-bit word machine the choices of $m = 2^{35} - 31$ and $a = 5^5$ appear to work well.)

Another generator of pseudo-random numbers uses recursions of the type

$$x_n = (ax_{n-1} + c) \quad \text{modulo } m$$

Such generators are called mixed congruential generators (as they involve both an additive and a multiplicative term). When using generators of this type, one often chooses m to equal the computer's word length, since this makes the computation of $(ax_{n-1} + c)$ modulo m—that is, the division of $ax_{n-1} + c$ by m—quite efficient.

Most computer languages already have a built-in random number generator which can be called to generate random numbers. For instance, BASIC utilizes two commands:

```
1  RANDOMIZE
2  U = RND
```

The result of the RANDOMIZE instruction is the request for the user to input the seed x_0. Once this is done, the quantity RND will represent the next pseudo-random number. For instance, the following program prints out the values of 10 pseudo-random numbers.

```
10  RANDOMIZE
20  FOR I = 1 TO 10
30  U = RND
40  PRINT U
50  NEXT
```

As our starting point in the computer simulation of systems we suppose that we can generate a sequence of pseudo-random numbers which can be taken as an approximation to the values of a sequence of independent uniform (0, 1) random variables. That is, we do not explore the interesting theoretical questions, which involve material outside the scope of this text, relating to the construction of "good" pseudo-random number generators. Rather, we assume that we have a "black box" that gives a random number on request.

3.2. Using Random Numbers to Evaluate Integrals

One of the earliest applications of random numbers was in the computation of integrals. Let $g(x)$ be a function and suppose we wanted to compute θ where

$$\theta = \int_0^1 g(x)\,dx$$

To compute the value of θ, note that if U is uniformly distributed over (0, 1) then we can express θ as

$$\theta = E[g(U)]$$

If $U_1, \ldots, U_k$ are independent uniform (0, 1) random variables, it thus follows that the random variables $g(U_1), \ldots g(U_k)$ are independent and identically distributed random variables having mean, θ. Therefore, by the strong law of large numbers, it follows that, with probability 1,

$$\sum_{i=1}^{k} \frac{g(U_i)}{k} \rightarrow E[g(U)] = \theta \qquad \text{as } k \rightarrow \infty$$

Hence we can approximate θ by generating a large number of random numbers u_i and taking as our approximation the average value of $g(u_i)$. This approach to approximating integrals is called the *Monte Carlo* approach.

The following BASIC algorithm can be used to approximate θ.

```
10 RANDOMIZE
20 INPUT K
30 S = 0
40 FOR I = 1 TO K
50 U = RND
60 S = S + g(U)
70 NEXT
80 PRINT S/K
```

The value printed is the approximation to $\theta = \int_0^1 g(x)\,dx$.

If we wanted to compute

$$\theta = \int_a^b g(x)\, dx$$

then, by making the substitution $y = (x - a)/(b - a)$, $dy = dx/(b - a)$, we see that

$$\theta = \int_0^1 g(a + [b - a]y)\,(b - a)\, dy$$

$$= \int_0^1 h(y)\, dy$$

where $h(y) = (b - a)g(a + [b - a]y)$. Thus we can approximate θ by continually generating random numbers and then taking the average value of h evaluated at these random numbers.

Similarly, if we wanted

$$\theta = \int_0^\infty g(x)\, dx$$

we could apply the substitution $y = 1/(x + 1)$, $dy = -dx/(x + 1)^2 = -y^2\, dx$, to obtain the identity

$$\theta = \int_0^1 h(y)\, dy$$

where

$$h(y) = \frac{g\left(\dfrac{1}{y} - 1\right)}{y^2}$$

The utility of using random numbers to approximate integrals becomes more apparent in the case of multidimensional integrals. Suppose that g is a function with an n-dimensional argument and that we are interested in computing

$$\theta = \int_0^1 \int_0^1 \cdots \int_0^1 g(x_1, \ldots, x_n)\, dx_1\, dx_2 \cdots dx_n$$

The key to the Monte Carlo approach to estimate θ lies in the fact that θ can be expressed as the following expectation:

$$\theta = E[g(U_1, \ldots, U_n)]$$

where $U_1, \ldots, U_n$ are independent uniform (0, 1) random variables. Hence, if we generate k independent sets, each consisting of n independent uniform (0, 1) random variables

$$\begin{gathered} U_1^1, \ldots, U_n^1 \\ U_1^2, \ldots, U_n^2 \\ \vdots \\ U_1^k, \ldots, U_n^k \end{gathered}$$

then, since the random variables $g(U_1^i, \ldots, U_n^i)$, $i = 1, \ldots, k$, are all independent and identically distributed random variables with mean θ, we can estimate θ by $\sum_{i=1}^k g(U_1^i, \ldots, U_n^i)/k$.

For an application of the above, consider the following approach to estimating π.

Example 3a: The Estimation of π

Suppose that the random vector (X, Y) is uniformly distributed in the square of area 4 centered at the origin. That is, it is a random point in the following region specified in Figure 3.1. Let us consider now the probability that this random point in the square is contained within the inscribed circle of radius 1 (see Figure 3.2). Note that since (X, Y) is uniformly distributed in the square it follows that

$$\begin{aligned} P\{(X, Y) \text{ is in the circle}\} &= P\{X^2 + Y^2 \le 1\} \\ &= \frac{\text{Area of the circle}}{\text{Area of the square}} = \frac{\pi}{4} \end{aligned}$$

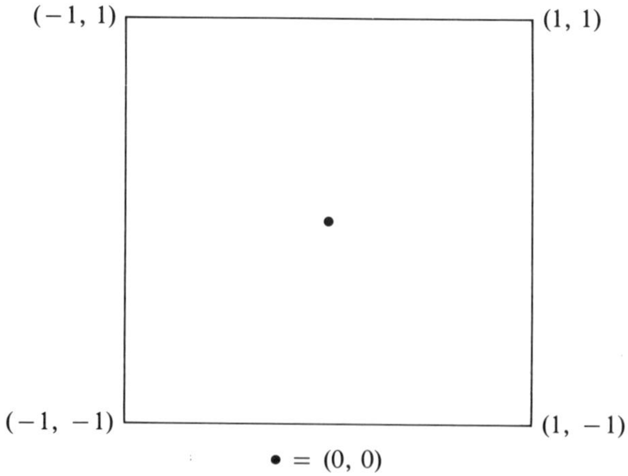

Figure 3.1

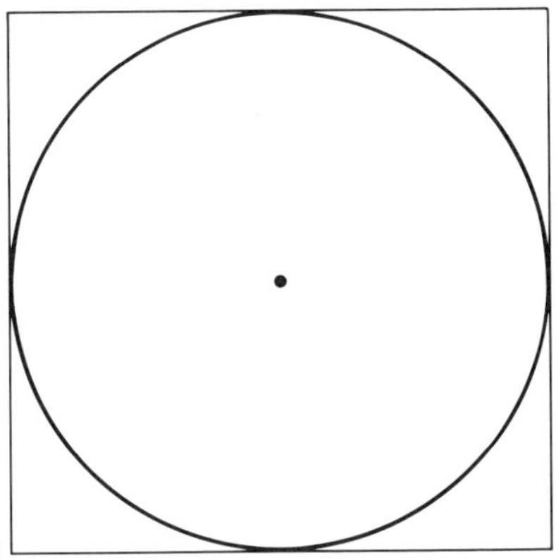

Figure 3.2

Hence, if we generate a large number of random points in the square, the proportion of points that fall within the circle will be approximately $\pi/4$. Now if X and Y were independent and both were uniformly distributed over $(-1, 1)$, their joint density would be

$$\begin{aligned} f(x, y) &= f(x)f(y) \\ &= \tfrac{1}{2} \cdot \tfrac{1}{2} \\ &= \tfrac{1}{4}, \qquad -1 \leq x \leq 1, \quad -1 \leq y \leq 1 \end{aligned}$$

Since the density function of (X, Y) is constant in the square, it thus follows (by definition) that (X, Y) is uniformly distributed in the square. Now if U is uniform on $(0, 1)$ then $2U$ is uniform on $(0, 2)$, and so $2U - 1$ is uniform on $(-1, 1)$. Therefore, if we generate random numbers U_1 and U_2, set $X = 2U_1 - 1$ and $Y = 2U_2 - 1$, and define

$$I = \begin{cases} 1 & \text{if } X^2 + Y^2 \leq 1 \\ 0 & \text{otherwise} \end{cases}$$

then

$$E[I] = P\{X^2 + Y^2 \leq 1\} = \frac{\pi}{4}$$

Hence we can estimate $\pi/4$ by generating a large number of pairs of random numbers u_1, u_2 and estimating $\pi/4$ by the fraction of pairs for which $(2u_1 - 1)^2 + (2u_2 - 1)^2 \leq 1$. ■

Thus random number generators can be used to generate the values of uniform (0, 1) random variables. Starting with these random numbers we show in Chapters 4 and 5 how we can generate the values of random variables from arbitrary distributions. With this ability to generate arbitrary random variables we will be able to simulate a probability system—that is, we will be able to generate, according to the specified probability laws of the system, all the random quantities of this system as it evolves over time.

Problems

1. If $x_0 = 5$ and

$$x_n = 3x_{n-1} \mod 150$$

find $x_1, \ldots, x_{10}$.

2. If $x_0 = 3$ and

$$x_n = 5x_{n-1} + 7 \mod 200$$

find $x_1, \ldots, x_{10}$.

In Problems 3–9 use simulation to approximate the following integrals. Compare your estimate with the exact answer if known.

3. $\int_0^1 \exp(e^x)\, dx$.

4. $\int_0^5 (1 - x^2)^{3/2}\, dx$.

5. $\int_{-5}^5 e^{x+x^2}\, dx$.

6. $\int_0^\infty x(1 + x)^{-2}\, dx$.

7. $\int_{-\infty}^\infty e^{-x^2}\, dx$.

8. $\int_0^1 \int_0^1 e^{(x+y)^2}\, dy\, dx$.

9. $\int_0^\infty \int_0^x e^{-(x+y)}\, dy\, dx$.

[Hint: Let $I_y(x) = \begin{cases} 1 & \text{if } y < x \\ 0 & \text{if } y \geq x \end{cases}$ and use this function to equate the integral to one in which both terms go from 0 to ∞.]

10. Use simulation to approximate $\text{Cov}(U, e^U)$, where U is uniform on (0, 1). Compare your approximation with the exact answer.

11. Let U be uniform on (0, 1). Use simulation to approximate the following.
(a) $\text{Cov}(U, \sqrt{1 - U^2})$.
(b) $\text{Cov}(U^2, \sqrt{1 - U^2})$.

12. For uniform (0, 1) random variables $U_1, U_2, \ldots$ define

$$N = \text{Minimum}\left\{n: \sum_{i=1}^{n} U_i > 1\right\}$$

That is, N is equal to the number of random numbers that must be summed to exceed 1.

(a) Estimate $E[N]$ by generating 100 values of N.
(b) Estimate $E[N]$ by generating 1000 values of N.
(c) Estimate $E[N]$ by generating 10000 values of N.
(d) What do you think is the value of $E[N]$?

13. Let U_i, $i \geq 1$, be random numbers. Define N by

$$N = \text{Maximum}\left\{n: \prod_{i=1}^{n} U_i \geq e^{-3}\right\}$$

where $\prod_{i=1}^{0} U_i \equiv 1$.

(a) Find $E[N]$ by simulation.
(b) Find $P\{N = i\}$, for $i = 0, 1, 2, 3, 4, 5, 6$, by simulation.

CHAPTER 4

Generating Discrete Random Variables

4.1. The Inverse Transform Method

Suppose we want to generate the value of a discrete random variable X having probability mass function

$$P\{X = x_j\} = p_j, \qquad j = 0, 1, \ldots, \quad \sum_j p_j = 1$$

To accomplish this, we generate a random number U—that is, U is uniformly distributed over (0, 1)—and set

$$X = \begin{cases} x_0 & \text{if} \quad U < p_0 \\ x_1 & \text{if } p_0 \leq U < p_0 + p_1 \\ \vdots & \\ x_j & \text{if } \sum_{i=1}^{j-1} p_i \leq U < \sum_{i=1}^{j} p_i \\ \vdots & \end{cases}$$

Since, for $0 < a < b < 1$, $P\{a \leq U < b\} = b - a$, we have that

$$P\{X = x_j\} = P\left\{\sum_{i=1}^{j-1} p_i \leq U < \sum_{i=1}^{j} p_i\right\} = p_j$$

and so X has the desired distribution.

REMARKS

1. The above can be written algorithmically as

Generate a random number U
If $U < p_0$ set $X = x_0$ and stop
If $U < p_0 + p_1$ set $X = x_1$ and stop
If $U < p_0 + p_1 + p_2$ set $X = x_2$ and stop
$\vdots$

2. If the x_i, $i \geq 0$, are ordered so that $x_0 < x_1 < x_2 < \cdots$ and if we let F denote the distribution function of X, then $F(x_k) = \Sigma_{i=0}^{k} p_i$ and so

$$X \text{ will equal } x_j \qquad \text{if} \qquad F(x_{j-1}) \leq U < F(x_j)$$

In other words, after generating a random number U we determine the value of X by finding the interval $(F(x_{j-1}), F(x_j))$ in which U lies [or, equivalently, by finding the inverse of $F(U)$]. It is for this reason that the above is called the discrete inverse transform method for generating X. ■

The amount of time it takes to generate a discrete random variable by the above method is proportional to the number of intervals one must search. For this reason it is sometimes worthwhile to consider the possible values x_j of X in decreasing order of the p_j.

Example 4a If we wanted to simulate a random variable X such that

$$p_1 = .20, \quad p_2 = .15, \quad p_3 = .25, \quad p_4 = .40 \qquad \text{where } p_j = P\{X = j\}$$

then we could generate U and do the following:

If $U < .20$ set $X = 1$ and stop
If $U < .35$ set $X = 2$ and stop
If $U < .60$ set $X = 3$ and stop
Otherwise set $X = 4$

However, a more efficient procedure is the following:

If $U < .40$ set $X = 4$ and stop
If $U < .65$ set $X = 3$ and stop
If $U < .85$ set $X = 1$ and stop
Otherwise set $X = 2$ ■

One case where it is not necessary to search for the appropriate interval in which the random number lies is when the desired random variable is the discrete uniform random variable. That is, suppose we want to generate the value of X which is equally likely to take on any of the values $1, \ldots, n$. That is, $P\{X = j\} = 1/n$, $j = 1, \ldots, n$. Using the above results it follows that we can accomplish this by generating U and then setting

$$X = j \qquad \text{if} \qquad \frac{j-1}{n} \leq U < \frac{j}{n}$$

Therefore X will equal j if $j - 1 \leq nU < j$; or, in other words,

$$X = \text{Int}(nU) + 1$$

where Int(x)—sometimes written as $[x]$—is the integer part of x (i.e., the largest integer less than or equal to x).

Discrete uniform random variables are quite important in simulation, as is indicated in the following two examples.

Example 4b: Generating a Random Permutation. Suppose we are interested in generating a permutation of the numbers $1, 2, \ldots, n$ which is such that all $n!$ possible orderings are equally likely. The following algorithm will accomplish this by first choosing one of the numbers $1, \ldots, n$ at random and then putting that number in position n; it then chooses at random one of the remaining $n - 1$ numbers and puts that number in position $n - 1$; it then chooses at random one of the remaining $n - 2$ numbers and puts it in position $n - 2$; and so on (where choosing a number at random means that each of the remaining numbers is equally likely to be chosen). However, so that we do not have to consider exactly which of the numbers remain to be positioned, it is convenient and efficient to keep the numbers in an ordered list and then randomly choose the position of the number rather than the number itself. That is, starting with any initial ordering $P_1, P_2, \ldots, P_n$ we pick one of the positions $1, \ldots, n$ at random and then interchange the number in that position with the one in position n. Now we randomly choose one of the positions $1, \ldots, n - 1$ and interchange the number in this position with the one in position $n - 1$, and so on.

Recalling that $\text{Int}(kU) + 1$ will be equally likely to take on any of the values $1, 2, \ldots, k$, we see that the above algorithm for generating a random permutation can be written as follows:

STEP 1: Let $P_1, P_2, \ldots, P_n$ be any permutation of $1, 2, \ldots, n$ (e.g., we can choose $P_j = j$, $j = 1, \ldots, n$).
STEP 2: Set $k = n$.
STEP 3: Generate a random number U and let $I = \text{Int}(kU) + 1$.
STEP 4: Interchange the values of P_I and P_k.
STEP 5: Let $k = k - 1$ and if $k > 1$ go to Step 3.
STEP 6: $P_1, \ldots, P_n$ is the desired random permutation.

For instance, suppose $n = 4$ and the initial permutation is 1, 2, 3, 4. If the first value of I (which is equally likely to be either 1, 2, 3, or 4) is $I = 3$, then the elements in positions 3 and 4 are interchanged and so the new permutation is 1, 2, 4, 3. If the next value of I is $I = 2$, then the elements in positions 2 and 3 are interchanged and so the new permutation is 1, 4, 2, 3. If the final value of I is $I = 2$, then the final permutation is 1, 4, 2, 3, and this is the value of the random permutation.

One very important property of the above algorithm is that it can also be used to generate a random subset, say of size r, of the integers $1, \ldots, n$. Namely, just follow the algorithm until the positions $n, n-1, \ldots, n-r+1$ are filled. The elements in these positions constitute the random subset. (In doing this we can always suppose that $r \leq n/2$; for if $r > n/2$ then we could choose a random subset of size $n - r$ and let the elements not in this subset be the random subset of size r.)

It should be noted that the ability to generate a random subset is particularly important in medical trials. For instance, suppose that a medical center is planning to test a new drug designed to reduce its user's blood cholesterol level. To test its effectiveness, the medical center has recruited 1000 volunteers to be subjects in the test. To take into account the possibility that the subjects' blood cholesterol levels may be affected by factors external to the test (such as changing weather conditions), it has been decided to split the volunteers into two groups of size 500—a *treatment* group that will be given the drug and a *control* that will be given a placebo. Both the volunteers and the administrators of the drug will not be told who is in each group (such a test is called *double-blind*). It remains to determine which of the volunteers should be chosen to constitute the treatment group. Clearly, one would want the treatment group and the control group to be as similar as possible in all respects with the exception that members in the first group are to receive the drug while those in the other group receive a placebo; for then it would be possible to conclude that any difference in response between the groups is indeed due to the drug. There is general agreement that the best way to accomplish this is to choose the 500 volunteers to be in the treatment group in a completely random fashion. That is, the choice should be made so that each of the $\binom{1000}{500}$ subsets of 500 volunteers is equally likely to constitute the set of volunteers. ■

Example 4c: Calculating Averages. Suppose we want to approximate $\bar{a} = \sum_{i=1}^{n} a(i)/n$, where n is large and the values $a(i)$, $i = 1, \ldots, n$, are complicated and not easily calculated. One way to accomplish this is to note that if X is a discrete uniform random variable over the integers $1, \ldots, n$, then the random variable $a(X)$ has a mean given by

$$E[a(X)] = \sum_{i=1}^{n} a(i)P\{X = i\} = \sum_{i=1}^{n} \frac{a(i)}{n} = \bar{a}$$

Hence, if we generate k discrete uniform random variables X_i, $i = 1, \ldots, k$—by generating k random numbers U_i and setting $X_i = \text{Int}(nU_i) + 1$—then each of the k random variables $a(X_i)$ will have mean $\bar{a}$ and so by the strong law of large numbers it follows that when k is large (though much smaller than n) the average of these values should approximately equal $\bar{a}$. Hence we can approximate $\bar{a}$ by using

$$\bar{a} \approx \sum_{i=1}^{k} \frac{a(X_i)}{k}$$ ■

Another random variable that can be generated without needing to search for the relevant interval in which the random number falls is the geometric.

Example 4d Recall that X is said to be a geometric random variable with parameter p if

$$P\{X = i\} = pq^{i-1}, \qquad i \geq 1, \text{ where } q = 1 - p$$

X can be thought of as representing the time of the first success when independent trials, each of which is a success with probability p, are performed. Since

$$\begin{aligned}\sum_{i=1}^{j-1} P\{X = i\} &= 1 - P\{X > j - 1\} \\ &= 1 - P\{\text{first } j - 1 \text{ trials are all failures}\} \\ &= 1 - q^{j-1}, \qquad j \geq 1\end{aligned}$$

we can generate the value of X by generating a random number U and setting X equal to that value j for which

$$1 - q^{j-1} \leq U < 1 - q^j$$

or, equivalently, for which

$$q^j < 1 - U \leq q^{j-1}$$

That is, we can define X by

$$X = \text{Min}\{j\colon q^j < 1 - U\}$$

Hence, using the fact that the logarithm is a monotone function, and so $a < b$ is equivalent to $\log(a) < \log(b)$, we obtain that X can be expressed as

$$\begin{aligned}X &= \text{Min}\{j\colon j \log(q) < \log(1 - U)\} \\ &= \text{Min}\left\{j\colon j > \frac{\log(1 - U)}{\log(q)}\right\}\end{aligned}$$

where the last inequality changed sign since $\log(q)$ is negative since $0 < q < 1$. Hence, using Int( ) notation we can express X as

$$X = \text{Int}\left(\frac{\log(1 - U)}{\log(q)}\right) + 1$$

Finally, by noting that $1 - U$ is also uniformly distributed on (0, 1), it follows that

$$X \equiv \operatorname{Int}\left(\frac{\log(U)}{\log(q)}\right) + 1$$

is also geometric with parameter p. ■

4.2. Generating a Poisson Random Variable

The random variable X is Poisson with mean λ if

$$p_i = P\{X = i\} = e^{-\lambda}\frac{\lambda^i}{i!} \qquad i = 0, 1, \ldots$$

The key to using the inverse transform method to generate such a random variable is the following identity (proved in Section 2.4 of Chapter 2):

$$p_{i+1} = \frac{\lambda}{i+1} p_i, \qquad i \geq 0 \tag{4.1}$$

Upon using the above recursion to compute the Poisson probabilities as they become needed, the inverse transform algorithm for generating a Poisson random variable with mean λ can be expressed as follows. (The quantity i refers to the value presently under consideration; $p = p_i$ is the probability that X equals i, and $F = F(i)$ is the probability that X is less than or equal to i.)

STEP 1: Generate a random number U.
STEP 2: $i = 0$, $p = e^{-\lambda}$, $F = p$.
STEP 3: If $U < F$, set $X = i$ and stop.
STEP 4: $p = \lambda p/(i + 1)$, $F = F + p$, $i = i + 1$.
STEP 5: Go to Step 3.

(In the above it should be noted that when we write, for example, $i = i + 1$, we do not mean that i is equal to $i + 1$ but rather that the value of i should be increased by 1.) To see that the above algorithm does indeed generate a Poisson random variable with mean λ, note that it first generates a random number U and then checks whether or not $U < e^{-\lambda} = p_0$. If so, it sets $X = 0$. If not, then it computes (in Step 4) p_1 by using the recursion (4.1). It now checks whether $U < p_0 + p_1$ (where the right-hand side is the new value of F), and if so it sets $X = 1$; and so on.

The above algorithm successively checks whether the Poisson value is 0 then whether it is 1, then 2, and so on. Thus the number of comparisons needed will be 1 greater than the generated value of the Poisson. Hence, on average, the above will need to make $1 + \lambda$ searches. Whereas this is fine when λ is small, it can be greatly improved upon when λ is large. Indeed, since a Poisson random variable with mean λ is most likely to take on one of the two integral values closest to λ, a more efficient algorithm would first check one of these values, rather than starting at 0 and working upward. Algorithm 4-1, written in BASIC and presented in the Appendix of Programs, lets $I = \text{Int}(\lambda)$ and computes p_I (by first taking logarithms and then by raising the result to the power e). It then uses Equation (4.1) to recursively determine $F(I)$. It now generates a Poisson random variable with mean λ by generating a random number U, and then noting whether or not $X \leq I$ by seeing whether or not $U \leq F(I)$. It then searches downward starting from $X = I$ in the case where $X \leq I$ and upward starting from $X = I + 1$ otherwise. (Algorithm 4-1 also avoids the potential difficulty that, due to round-off error, when λ is large the computer might take $e^{-\lambda}$ to equal 0.)

REMARK

The number of searches needed by Algorithm 4-1 is roughly 1 more than the absolute difference between the random variable X and its mean λ. Since for λ large a Poisson is (by the central limit theorem) approximately normal with mean and variance both equal to λ, it follows that

$$\begin{aligned}
\text{Average number of searches} &\simeq 1 + E[|X - \lambda|] \qquad \text{where } X \sim N(\lambda, \lambda)^{\dagger} \\
&= 1 + \sqrt{\lambda}\, E\left[\frac{|X - \lambda|}{\sqrt{\lambda}}\right] \\
&= 1 + \sqrt{\lambda}\, E[|Z|] \qquad \text{where } Z \sim N(0, 1) \\
&\simeq 1 + .82\sqrt{\lambda} \qquad \text{(see Problem 11)}
\end{aligned}$$

That is, using Algorithm 4-1, the average number of searches grows with the square root of λ rather than with λ as λ becomes larger and larger. ■

4.3. Generating Binomial Random Variables

Suppose we want to generate the value of a binomial (n, p) random variable X—that is, X is such that

$$P\{X = i\} = \frac{n!}{i!(n - i)!} p^i(1 - p)^{n-i}, \qquad i = 0, 1, \ldots, n$$

To do so, we employ the inverse transform method by making use of the recursive identity

$$P\{X = i + 1\} = \frac{n - i}{i + 1} \frac{p}{1 - p} P\{X = i\}$$

† We use the notation $X \sim F$ to mean that X has distribution function F. The symbol $N(\mu, \sigma^2)$ stands for a normal distribution with mean μ and variance σ^2.

With i denoting the value currently under consideration, pr $= P\{X = i\}$ the probability that X is equal to i, and $F = F(i)$ the probability that X is less than or equal to i, the algorithm can be expressed as follows:

INVERSE TRANSFORM ALGORITHM FOR GENERATING A BINOMIAL (n, p) RANDOM VARIABLE

STEP 1: Generate a random number U.
STEP 2: $c = p/(1 - p)$, $i = 0$, $\text{pr} = (1 - p)^n$, $F = \text{pr}$.
STEP 3: If $U < F$, set $X = i$ and stop.
STEP 4: $\text{pr} = [c(n - i)/(i + 1)]\text{pr}$, $F = F + \text{pr}$, $i = i + 1$.
STEP 5: Go to Step 3.

The above algorithm first checks whether $X = 0$, then whether $X = 1$, and so on. Hence the number of searches it makes is 1 more than the value of X. Therefore, on average, it will take $1 + np$ searches to generate X. By recalling that a binomial (n, p) random variable represents the number of successes in n independent trials when each is a success with probability p, it follows that such a random variable can also be generated by subtracting from n the value of a binomial $(n, 1 - p)$ random variable (why is that?). Hence, when $p > \frac{1}{2}$, we can generate a binomial $(n, 1 - p)$ random variable by the above method and subtract its value from n to obtain the desired generation.

REMARKS

1. Another way of generating a binomial (n, p) random variable is by utilizing its interpretation as the number of successes in n independent trials by generating n random numbers $U_1, \ldots, U_n$ and then setting X equal to the number of the U_i that are less than or equal to p. By regarding the ith trial as a success if $U_i < p$ and noting that the probability of this event is equal to p, it is easy to see that this results in a binomial (n, p) random variable. However, this approach requires n random numbers and makes n comparisons, whereas the above algorithm only requires 1 random number and makes, on average, $1 + np$ comparisons (along with an equal number of divisions).

2. As in the Poisson case, when the mean np is large it is better to first determine if the generated value is less than or equal to $I \equiv \text{Int}(np)$ or whether it is greater than I. In the former case one should then start the search with $X = I$, then $X = I - 1, \ldots,$ and so on; whereas in the latter case one should start searching with $I + 1$ and go upward. ■

4.4. The Acceptance–Rejection Technique

Suppose we have an efficient method for simulating a random variable having probability mass function $\{q_j, j \geq 0\}$. We can use this as the basis for simulating from the distribution having mass function $\{p_j, j \geq 0\}$ by first simulating a random variable Y having mass function $\{q_j\}$ and then accepting this simulated value with a probability proportional to p_Y/q_Y.

Specifically, let c be a constant such that

$$\frac{p_j}{q_j} \leq c \qquad \text{for all } j \text{ such that } p_j \neq 0 \tag{4.2}$$

We now have the following technique, called the rejection method, for simulating a random variable X having mass function $p_j = P\{X = j\}$.

Rejection Method

STEP 1: Simulate the value of Y, having probability mass function q_j.
STEP 2: Generate a random number U.
STEP 3: If $U < p_Y/cq_Y$, set $X = Y$ and stop. Otherwise return to Step 1.

The rejection method is pictorially represented in Figure 4.1.

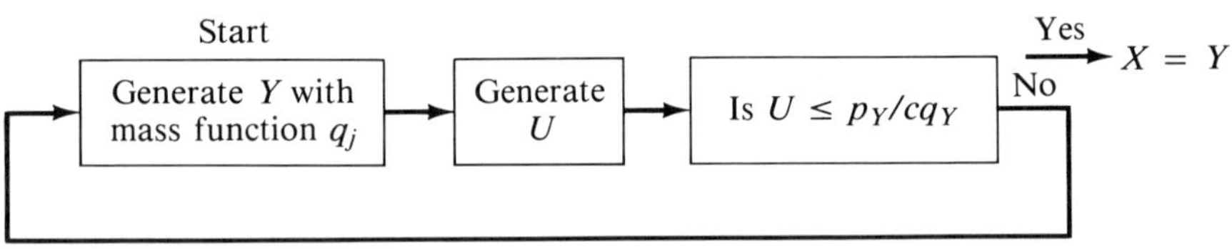

Figure 4.1 Acceptance–rejection.

We now prove that the rejection method works.

Theorem:

The random variable X obtained from the rejection method has probability mass function $P\{X = i\} = p_i$.

Proof.

$$\begin{aligned} P\{X = i\} &= P\{Y = i \mid \text{Acceptance}\} \\ &= \frac{P\{Y = i, \text{ Acceptance}\}}{K}, \qquad \text{where } K = P\{\text{Acceptance}\} \\ &= \frac{P\left\{Y = i,\ U \leq \dfrac{p_i}{cq_i}\right\}}{K} \end{aligned}$$

$$= \frac{P\{Y = i\}\, P\left\{U \leq \frac{p_i}{cq_i}\right\}}{K} \qquad \text{by independence}$$

$$= \frac{q_i}{K} \frac{p_i}{cq_i}$$

$$= \frac{p_i}{cK} \tag{4.3}$$

Summing over all i yields that

$$1 = \sum_i P\{X = i\} = \frac{1}{cK} \sum_i p_i$$
$$= \frac{1}{cK}$$

and so we see from (4.3) that the generated random variable X indeed has the desired mass function. □

REMARKS

1. The reader should note that the way in which we "accept the value Y with probability p_Y/cq_Y" is by generating a random number U and then accepting Y if $U \leq p_Y/cq_Y$.

2. Since each iteration of the algorithm will, independently, result in an accepted value with probability $P\{\text{Acceptance}\} = K = 1/c$, it follows that the number of iterations needed will have a geometric distribution with parameter $1/c$ and will thus have mean c. Therefore the closer c is to 1, and it can never be less than 1 [which can be seen by writing the inequality (4.2) as $p_i \leq cq_i$ and then summing both sides], the more efficient is this approach; which is intuitively clear since the closer c is to 1 the more alike are the two mass functions $\{p_j\}$ and $\{q_j\}$. ■

Example 4e Suppose we wanted to simulate the value of a random variable X that takes one of the values 1, 2, . . . , 10 with respective probabilities .11, .12, .09, .08, .12, .10, .09, .09, .10, .10. Whereas one possibility is to use the inverse transform algorithm, a better approach is to use the rejection method with q being the discrete uniform density on 1, . . . , 10. That is, $q_j = 1/10, j = 1, \ldots, 10$. For this choice of $\{q_j\}$ we can choose c by

$$c = \text{Max} \frac{p_j}{q_j} = 1.2$$

and so the algorithm would be as follows:

STEP 1: Generate a random number U_1 and set $Y = \text{Int}(10U_1) + 1$.
STEP 2: Generate a second random under U_2.
STEP 3: If $U_2 \leq p_Y/.12$, set $X = Y$ and stop. Otherwise return to Step 1.

The constant .12 in Step 3 arises since $cq_Y = 1.2/10 = .12$. On average, this algorithm requires only 1.2 iterations of the above to obtain the generated value of X. ■

The power of the rejection method, a version of which was initially proposed by the famous mathematician John von Newmann, will become even more readily apparent when we consider its analogue when generating continuous random variables.

4.5. The Composition Approach

Suppose that we had an efficient method to simulate the value of a random variable having either of the two probability mass functions $\{p_j, j \geq 0\}$ or $\{q_j, j \geq 0\}$, and suppose we wanted to simulate the value of the random variable X having mass function

$$P\{X = j\} = \alpha p_j + (1 - \alpha)q_j, \qquad j \geq 0 \tag{4.4}$$

where $0 < \alpha < 1$. One way to simulate such a random variable X is to note that if X_1 and X_2 are random variables having respective mass functions $\{p_j\}$ and $\{q_j\}$, then the random variable X defined by

$$X = \begin{cases} X_1 & \text{with probability } \alpha \\ X_2 & \text{with probability } 1 - \alpha \end{cases}$$

will have its mass function given by (4.4). From this it follows that we can generate the value of such a random variable by first generating a random number U and then generating a value of X_1 if $U < \alpha$ and of X_2 if $U > \alpha$.

Example 4f Suppose we want to generate the value of a random variable X such that

$$p_j = P\{X = j\} = \begin{cases} .05 & \text{for } j = 1, 2, 3, 4, 5 \\ .15 & \text{for } j = 6, 7, 8, 9, 10 \end{cases}$$

By noting that $p_j = .5p_j^1 + .5p_j^2$, where

$$p_j^1 = .1, \quad j = 1, \ldots, 10 \qquad \text{and} \qquad p_j^2 = \begin{cases} 0 & \text{for } j = 1, 2, 3, 4, 5 \\ .2 & \text{for } j = 6, 7, 8, 9, 10 \end{cases}$$

we can accomplish this by first generating a random number U and then generating from the discrete uniform over $1, \ldots, 10$ if $U < .5$ and from the discrete uniform over 6, 7, 8, 9, 10 otherwise. That is, we can simulate X as follows:

STEP 1: Generate a random number U_1.
STEP 2: Generate a random number U_2.
STEP 3: If $U_1 < .5$, set $X = \text{Int}(10U_2) + 1$. Otherwise set $X = \text{Int}(5U_2) + 6$. ■

If F_i, $i = 1, \ldots, n$, are distribution functions and α_1, $i = 1, \ldots, n$, are nonnegative numbers summing to 1, then the distribution function F given by

$$F(x) = \sum_{i=1}^{n} \alpha_i F_i(x)$$

is said to be a *mixture*, or a *composition*, of the distribution functions F_i, $i = 1, \ldots, n$. One way to simulate from F is first to simulate a random variable I, equal to i with probability α_i, $i = 1, \ldots, n$, and then to simulate from the distribution F_I. (That is, if the simulated value of I is $I = j$, then the second simulation is from F_j.) This approach to simulating from F is often referred to as the *composition method.*

Problems

1. Write a program to generate n values from the probability mass function $p_1 = \frac{1}{3}$, $p_2 = \frac{2}{3}$.
(a) Let $n = 100$, run the program and determine the proportion of values that are equal to 1.
(b) Repeat (a) with $n = 1000$.
(c) Repeat (a) with $n = 10{,}000$.

2. Write a computer program that, when given a probability mass function $\{p_j, j = 1, \ldots, n\}$ as an input, gives as an output the value of a random variable having this mass function.

3. Give an efficient algorithm to simulate the value of a random variable X such that

$$P\{X = 1\} = .3, \quad P\{X = 2\} = .2, \quad P\{X = 3\} = .35, \quad P\{X = 4\} = .15$$

4. A deck of 100 cards—numbered $1, 2, \ldots, 100$—are shuffled and then turned over one at a time. Say that a "hit" occurs whenever card i is the ith card to be turned over, $i = 1, \ldots, 100$. Write a simulation program to estimate the expectation and variance of the total number of hits. Run the program. Find the exact answers and compare them with your estimates.

5. Another method of generating a random permutation, different from the one presented in Example 4b, is to successively generate a random permutation of the elements $1, 2, \ldots, n$ starting with $n = 1$, then $n = 2$, and so on. (Of course, the random permutation when $n = 1$ is 1). Once one has a random permutation of the first $n - 1$ elements—call it $P_1, \ldots, P_{n-1}$—the random permutation of the n elements $1, \ldots, n$ is obtained by putting n in the final position—to obtain the permutation $P_1, \ldots, P_{n-1}, n$—and then interchanging the element in position n (namely, n) with the element in a randomly chosen position which is equally likely to be either position 1, position 2, . . . , or position n.

(a) Write an algorithm that accomplishes the above.

(b) Prove by mathematical induction on n that the algorithm works, in that the permutation obtained is equally likely to be any of the $n!$ permutations of $1, 2, \ldots, n$.

6. (a) Using 100 random numbers explain how you could find an approximation for $\Sigma_{i=1}^{N} e^{i/N}$, where $N = 10{,}000$.

(b) Obtain the approximation.

(c) Is your approximation a good one?

7. A pair of fair dice are to be continually rolled until all the possible outcomes $2, 3, \ldots, 12$ have occurred at least once. Develop a simulation study to estimate the expected number of dice rolls that are needed.

8. Suppose that each item on a list of n items has a value attached to it, and let $v(i)$ denote the value attached to the ith item on the list. Suppose that n is very large, and also that each item may appear at many different places on the list. Explain how random numbers can be used to estimate the sum of the values of the different items on the list (where the value of each item is to be counted once no matter how many times the item appears on the list).

9. Consider the n events $A_1, \ldots, A_n$ where the event A_i, $i = 1, \ldots, n$, consists of the following n_i outcomes: $A_i = \{a_{i,1}, a_{i,2}, \ldots, a_{i,n_i}\}$. Suppose that for any given outcome a, $P\{a\}$, the probability that the experiment results in outcome a, is known. Explain how one can use the results of Problem 8 to estimate $P\{U_{i=1}^{n} A_i\}$, the probability that at least one of the events A_i occurs. Note that the events A_i, $i = 1, \ldots, n$, are not assumed to be mutually exclusive.

10. The negative binomial probability mass function with parameters (r, p), where r is a positive integer and $0 < p < 1$, is given by

$$p_j = \frac{(j-1)!}{(j-r)!\,(r-1)!}\, p^r(1-p)^{j-r}, \qquad j = r, r+1, \ldots$$

(a) Use the relationship between negative binomial and geometric random variables and the results of Example 4d to obtain an algorithm for simulating from this distribution.
(b) Verify the relation

$$p_{j+1} = \frac{j(1-p)}{j+1-r} p_j$$

(c) Use the relation in part (b) to give a second algorithm for generating negative binomial random variables.
(d) Use the interpretation of the negative binomial distribution as the number of trials it takes to amass a total of r successes when each trial independently results in a success with probability p, to obtain still another approach for generating such a random variable.

11. If Z is a unit normal random variable, show that

$$E[|Z|] = \left(\frac{2}{\pi}\right)^{1/2} \approx .82$$

12. Give a method for simulating from the probability mass function p_j, $j = 5, 6, \ldots, 14$, where

$$p_j = \begin{cases} .11 & \text{when } j \text{ is odd and } 5 \leq j \leq 13 \\ .09 & \text{when } j \text{ is even and } 6 \leq j \leq 14 \end{cases}$$

13. Suppose that the random variable X can take on any of the values 1, . . . , 10 with respective probabilities .06, .06, .06, .06, .06, .15, .13, .14, .15, .13. Use the composition approach to give an algorithm that generates the value of X.

14. Present a method to generate the value of X, where

$$P\{X = j\} = (\tfrac{1}{2})^{j+1} + \frac{(\frac{1}{2})2^{j-1}}{3^j}, \qquad j = 1, 2, \ldots$$

15. Let X have mass function $p_j = P\{X = j\}$, $\Sigma_{j=1}^{\infty} p_j = 1$. Let

$$\lambda_n = P\{X = n | X > n - 1\} = \frac{p_n}{1 - \sum_{j=0}^{n-1} p_j}, \qquad n = 1, \ldots$$

(a) Show that $p_1 = \lambda_1$ and

$$p_n = (1 - \lambda_1)(1 - \lambda_2) \cdots (1 - \lambda_{n-1})\lambda_n$$

The quantities λ_n, $n \geq 1$, are called the discrete hazard rates since if we think of X as the lifetime of some item then λ_n represents the probability that an item that has reached the age n will die during that time period. The following approach to simulating discrete random variables, called the discrete hazard rate method, generates a succession of random numbers, stopping when the nth random number is less than λ_n. The algorithm can be written as follows:

STEP 1: $X = 1$
STEP 2: Generate a random number U
STEP 3: If $U < \lambda_X$ stop
STEP 4: $X = X + 1$
STEP 5: Go to Step 2

(b) Show that the value of X when the above stops has the desired mass function.
(c) Suppose that X is a geometric random variable with parameter p. Determine the values λ_n, $n \geq 1$. Explain what the above algorithm is doing in this case and why its validity is clear.

CHAPTER 5

Generating Continuous Random Variables

Introduction

Each of the techniques for generating a discrete random variable has its analog in the continuous case. In Sections 5.1 and 5.2 we present the inverse transform approach and the rejection approach for generating continuous random variables. In Section 5.3 we consider a powerful approach for generating normal random variables, known as the polar method. Finally, in Sections 5.4 and 5.5 we consider the problem of generating Poisson and nonhomogeneous Poisson processes.

5.1. The Inverse Transform Algorithm

Consider a continuous random variable having distribution function F. A general method for generating such a random variable—called the inverse transformation method—is based on the following proposition.

Proposition:

Let U be a uniform (0, 1) random variable. For any continuous distribution function F the random variable X defined by

$$X = F^{-1}(U)$$

has distribution F. [$F^{-1}(u)$ is defined to be that value of x such that $F(x) = u$.]

Proof. Let F_X denote the distribution function of $X = F^{-1}(U)$. Then

$$\begin{aligned} F_X(x) &= P\{X \le x\} \\ &= P\{F^{-1}(U) \le x\} \end{aligned} \tag{5.1}$$

Now since F is a distribution function it follows that $F(x)$ is a monotone increasing function of x and so the inequality "$a \le b$" is equivalent to the inequality "$F(a) \le F(b)$." Hence, from Equation (5.1), we see that

$$\begin{aligned} F_X(x) &= P\{F(F^{-1}(U) \le F(x)\} \\ &= P\{U \le F(x)\} \qquad \text{since } F(F^{-1}(U)) = U \\ &= F(x) \qquad \text{since } U \text{ is uniform } (0, 1) \end{aligned}$$ □

The above proposition thus shows that we can generate a random variable X from the continuous distribution function F by generating a random number U and then setting $X = F^{-1}(U)$.

Example 5a Suppose we wanted to generate a random variable X having distribution function

$$F(x) = x^n, \qquad 0 < x < 1$$

If we let $x = F^{-1}(u)$, then

$$u = F(x) = x^n \qquad \text{or, equivalently,} \qquad x = u^{1/n}$$

Hence we can generate such a random variable X by generating a random number U and then setting $X = U^{1/n}$. ■

The inverse transform method yields a powerful approach to generating exponential random variables, as is indicated in the next example.

Example 5b If X is an exponential random variable with rate 1, then its distribution function is given by

$$F(x) = 1 - e^{-x}$$

If we let $x = F^{-1}(u)$, then

$$u = F(x) = 1 - e^{-x}$$

or

$$1 - u = e^{-x}$$

or, taking logarithms,

$$x = -\log(1 - u)$$

Hence we can generate an exponential with parameter 1 by generating a random number U and then setting

$$X = F^{-1}(U) = -\log(1 - U)$$

A small savings in time can be obtained by noting that $1 - U$ is also uniform on (0, 1) and thus $-\log(1 - U)$ has the same distribution as $-\log U$. That is, the negative logarithm of a random number is exponentially distributed with rate 1.

In addition, note that if X is exponential with mean 1 then, for any positive c, cX is exponential with mean c. Hence an exponential random variable X with rate λ (mean $1/\lambda$) can be generated by generating a random number U and setting

$$X = -\frac{1}{\lambda} \log U \qquad \blacksquare$$

REMARK

The above also provides us with another algorithm for generating a Poisson random variable. To begin, recall that a Poisson process with rate λ results when the times between successive events are independent exponentials with rate λ. (See Section 2.5 of Chapter 2.) For such a process $N(1)$, the number of events by time 1, is Poisson distributed with mean λ. However, if we let X_i, $i = 1, \ldots,$ denote the successive interarrival times, then the nth event will occur at time $\sum_{i=1}^{n} X_i$, and so the number of events by time 1 can be expressed as

$$N(1) = \text{Max}\left\{n: \sum_{i=1}^{n} X_i \leq 1\right\}$$

That is, the number of events by time 1 is equal to the largest n for which the nth event has occurred by time 1. (For example, if the fourth event occurred by time 1 but the fifth event did not, then clearly there would have been a total of four events by time 1.) Hence, using the results of Example 5b, we can generate $N = N(1)$, a Poisson random variable with mean λ, by generating random numbers $U_1, \ldots, U_n, \ldots$ and setting

$$\begin{aligned} N &= \text{Max}\left\{n: \sum_{i=1}^{n} -\frac{1}{\lambda} \log U_i \leq 1\right\} \\ &= \text{Max}\left\{n: \sum_{i=1}^{n} \log U_i \geq -\lambda\right\} \\ &= \text{Max}\{n: \log(U_1 \cdots U_n) \geq -\lambda\} \\ &= \text{Max}\{n: U_1 \cdots U_n \geq e^{-\lambda}\} \end{aligned}$$

Hence a Poisson random variable N with mean λ can be generated by successively generating random numbers until their product falls below $e^{-\lambda}$, and then setting N equal to 1 less than the number of random numbers required. That is,

$$N = \text{Min}\{n: U_1 \cdots U_n < e^{-\lambda}\} - 1 \qquad \blacksquare$$

The results of Example 5b along with the relationship between the gamma and the exponential distribution can be used to efficiently generate a gamma (n, λ) random variable.

Example 5c Suppose we wanted to generate the value of a gamma (n, λ) random variable. Since the distribution function F of such a random variable is given by

$$F(x) = \int_0^x \frac{\lambda e^{-\lambda y}(\lambda y)^{n-1}}{(n-1)!}\, dy$$

it is not possible to give a closed form expression for its inverse. However, by using the result that a gamma (n, λ) random variable X can be regarded as being the sum of n independent exponentials, each with rate λ (see Section 2.3 of Chapter 2), we can make use of Example 5b to generate X. Specifically, we can generate a gamma (n, λ) random variable by generating n random numbers $U_1, \ldots, U_n$ and then setting

$$\begin{aligned} X &= -\frac{1}{\lambda}\log U_1 - \cdots - \frac{1}{\lambda}\log U_n \\ &= -\frac{1}{\lambda}\log (U_1 \cdots U_n) \end{aligned}$$

where the use of the identity $\Sigma_{i=1}^n \log x_i = \log(x_1 \cdots x_n)$ is computationally time saving in that it requires only one rather than n logarithm computations. ■

The results of Example 5c can be used to provide an efficient way of generating a set of exponential random variables by first generating their sum and then, conditional on the value of that sum, generating the individual values. For example, we could generate X and Y, a pair of independent and identically distributed exponentials having mean 1, by first generating $X + Y$ and then using the result (Problem 35 of Chapter 2) that, given that $X + Y = t$, the conditional distribution of X is uniform on $(0, t)$. The following algorithm can thus be used to generate a pair of exponentials with mean 1.

STEP 1: Generate random numbers U_1 and U_2.
STEP 2: Set $t = -\log(U_1 U_2)$.
STEP 3: Generate a random number U_3.
STEP 4: $X = tU_3$, $Y = t - X$.

Comparing the above with the more direct approach of generating two random numbers U_1 and U_2 and then setting $X = -\log U_1$, $Y = -\log U_2$ shows that the above algorithm saves a logarithmic computation at the cost of two multiplications and the generation of a random number.

We can also generate k independent exponentials with mean 1 by first generating their sum, say by $-\log (U_1 \cdots U_k)$, and then generating $k - 1$ additional random numbers $U_1, \cdots, U_{k-1}$, which should then be ordered. If $U_{(1)} < U_{(2)} < \cdots < U_{(k-1)}$ are their ordered values, and if $-\log (U_1 \cdots U_k) = t$, then the k exponentials are

$$t[U_{(i)} - U_{(i-1)}], \qquad i = 1, 2, \cdots, k, \text{ where } U_{(0)} \equiv 0,\ U_{(k)} \equiv t \qquad \blacksquare$$

5.2. The Rejection Method

Suppose we have a method for generating a random variable having density function $g(x)$. We can use this as the basis for generating from the continuous distribution having density function $f(x)$ by generating Y from g and then accepting this generated value with a probability proportional to $f(Y)/g(Y)$.

Specifically, let c be a constant such that

$$\frac{f(y)}{g(y)} \leq c \qquad \text{for all } y$$

We then have the following technique for generating a random variable having density f.

THE REJECTION METHOD

STEP 1: Generate Y having density g.
STEP 2: Generate a random number U.
STEP 3: If $U \leq \dfrac{f(Y)}{cg(Y)}$, set $X = Y$. Otherwise return to Step 1.

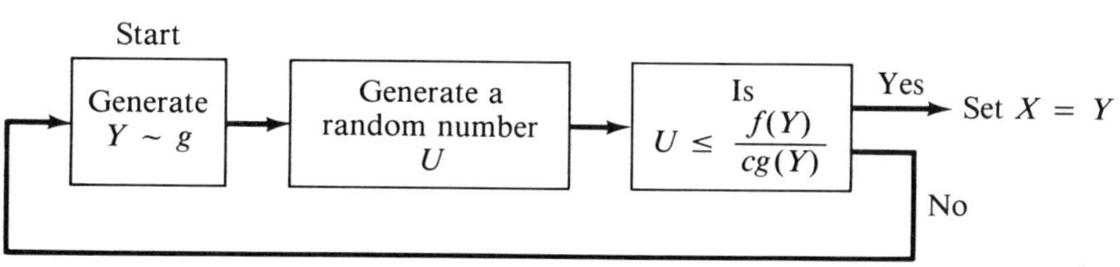

Figure 5.1 The rejection method for simulating a random variable X having density function f.

The reader should note that the rejection method is exactly the same as in the case of discrete random variables with the only difference being that densities replace mass functions. In exactly the same way as we did in the discrete case we can prove the following result.

Theorem:

(i) The random variable generated by the rejection method has density f.
(ii) The number of iterations of the algorithm that are needed is a geometric random variable with mean c.

As in the discrete case it should be noted that the way in which one accepts the value Y with probability $f(Y)/cg(Y)$ is by generating a random number U and then accepting Y if $U \leq f(Y)/cg(Y)$.

Example 5d Let us use the rejection method to generate a random variable having density function

$$f(x) = 20\,x(1 - x)^3, \qquad 0 < x < 1$$

Since this random variable (which is beta with parameters 2, 4) is concentrated in the interval (0, 1), let us consider the rejection method with

$$g(x) = 1, \qquad 0 < x < 1$$

To determine the constant c such that $f(x)/g(x) \leq c$, we use calculus to determine the maximum value of

$$\frac{f(x)}{g(x)} = 20\,x(1 - x)^3$$

Differentiation of this quantity yields

$$\frac{d}{dx}\left(\frac{f(x)}{g(x)}\right) = 20\,[(1 - x)^3 - 3x(1 - x)^2]$$

Setting this equal to 0 shows that the maximal value is attained when $x = \frac{1}{4}$ and thus

$$\frac{f(x)}{g(x)} \leq 20(\tfrac{1}{4})\,(\tfrac{3}{4})^3 = \tfrac{135}{64} \equiv c$$

Hence

$$\frac{f(x)}{cg(x)} = \frac{256}{27}\, x(1-x)^3$$

and thus the rejection procedure is as follows:

STEP 1: Generate random numbers U_1 and U_2.
STEP 2: If $U_2 \leq \frac{256}{27}\, U_1(1-U_1)^3$, stop and set $X = U_1$. Otherwise return to Step 1.

The average number of times that Step 1 will be performed is $c = \frac{135}{64} \approx 2.11$. ■

Example 5e Suppose we wanted to generate a random variable having the gamma $(\frac{3}{2}, 1)$ density

$$f(x) = K\, x^{1/2}\, e^{-x}, \qquad x > 0$$

where $K = 1/\Gamma(\frac{3}{2}) = 2/\sqrt{\pi}$. Because such a random variable is concentrated on the positive axis and has mean $\frac{3}{2}$, it is natural to try the rejection technique with an exponential random variable with the same mean. Hence let

$$g(x) = \frac{2e^{-2x/3}}{3}, \qquad x > 0$$

Now

$$\frac{f(x)}{g(x)} = \frac{3K}{2}\, x^{1/2}\, e^{-x/3}$$

By differentiating and setting the resultant derivative equal to 0, we obtain that the maximal value of this ratio is obtained when

$$\tfrac{1}{2}x^{-1/2}\, e^{-x/3} = \tfrac{1}{3}\, x^{1/2}\, e^{-x/3}$$

that is, when $x = \frac{3}{2}$. Hence

$$c = \text{Max}\, \frac{f(x)}{g(x)} = \frac{3K}{2}\, (\tfrac{3}{2})^{1/2}\, e^{-1/2}$$

$$= \frac{3^{3/2}}{(2\pi e)^{1/2}} \qquad \text{since } K = 2/\sqrt{\pi}$$

Since

$$\frac{f(x)}{cg(x)} = (2e/3)^{1/2}\, x^{1/2} e^{-x/3}$$

we see that a gamma $(\frac{3}{2}, 1)$ random variable can be generated as follows:

STEP 1: Generate a random number U_1 and set $Y = -\frac{3}{2} \log U_1$.
STEP 2: Generate a random number U_2.

STEP 3: If $U_2 < (2e/3)^{1/2}\, Y^{1/2}e^{-Y/3}$, set $X = Y$. Otherwise return to Step 1.

The average number of iterations that will be needed is

$$c = 3\left(\frac{3}{2\pi e}\right)^{1/2} \approx 1.257.$$

REMARK

Even though the gamma $(\frac{3}{2}, 1)$ random variable has mean $\frac{3}{2}$ it is not immediately apparent that we should use the rejection with an exponential with the same mean. Indeed, suppose that we let

$$g(x) = \lambda e^{-\lambda x}$$

Then

$$\frac{f(x)}{g(x)} = \frac{Kx^{1/2}\, e^{-(1-\lambda)x}}{\lambda}$$

The maximal value of this ratio is obtained when

$$\tfrac{1}{2}\, x^{-1/2} = (1 - \lambda)x^{1/2}$$

or when $x = [2(1 - \lambda)]^{-1}$ [provided $\lambda < 1$; if $\lambda \geq 1$, it is easy to see that the ratio $f(x)/g(x)$ assumes arbitrarily large values]. Hence, if we use the exponential with mean $1/\lambda$, then the average number of iterations of the algorithm that will be needed is

$$c = \text{Max}\,\frac{f(x)}{g(x)} = \frac{K}{\lambda}\,[2(1 - \lambda)]^{-1/2}\, e^{-1/2}$$

Thus the best choice of λ is the one that minimizes the above or, equivalently, that maximizes $\lambda(1 - \lambda)^{1/2}$. Calculus now shows that this value is such that

$$(1 - \lambda)^{1/2} = \frac{\lambda(1 - \lambda)^{-1/2}}{2}$$

or, equivalently,

$$1 - \lambda = \lambda/2 \qquad \text{or} \qquad \lambda = \tfrac{2}{3}$$

Hence the best exponential to use in the rejection method to generate a gamma $(\frac{3}{2}, 1)$ random variable is indeed the exponential with mean $\frac{3}{2}$. ■

Our next example shows how the rejection technique can be used to generate normal random variables.

Example 5f: Generating a Normal Random Variable. To generate a unit normal random variable Z (i.e., one with mean 0 and variance 1), note first that the absolute value of Z has probability density function

$$f(x) = \frac{2}{\sqrt{2\pi}} e^{-x^2/2} \qquad 0 < x < \infty \tag{5.2}$$

We start by generating from the preceding density function by using the rejection method with g being the exponential density function with mean 1—that is,

$$g(x) = e^{-x} \qquad 0 < x < \infty$$

Now

$$\frac{f(x)}{g(x)} = \sqrt{2/\pi}\, e^{x - x^2/2}$$

and so the maximum value of $f(x)/g(x)$ occurs at the value of x that maximizes $x - x^2/2$. Calculus shows that this occurs when $x = 1$ and so we can take

$$c = \text{Max}\,\frac{f(x)}{g(x)} = \frac{f(1)}{g(1)} = \sqrt{2e/\pi}$$

Because

$$\begin{aligned}\frac{f(x)}{cg(x)} &= \exp\left\{x - \frac{x^2}{2} - \frac{1}{2}\right\} \\ &= \exp\left\{-\frac{(x-1)^2}{2}\right\}\end{aligned}$$

it follows that we can generate the absolute value of a unit normal random variable as follows:

STEP 1: Generate Y, an exponential random variable with rate 1.
STEP 2: Generate a random number U.
STEP 3: If $U \leq \exp\{-(Y-1)^2/2\}$, set $X = Y$. Otherwise return to Step 1.

Once we have simulated a random variable X having density function as in Equation (5.2)—and such a random variable is thus distributed as the absolute value of a unit normal—we can then obtain a unit normal Z by letting Z be equally likely to be either X or $-X$.

In Step 3, the value Y is accepted if $U \leq \exp\{-(Y-1)^2/2\}$, which is equivalent to $-\log U \geq (Y-1)^2/2$. However, in Example 5b it was shown that $-\log U$ is exponential with rate 1, and so the above is equivalent to the following:

STEP 1: Generate independent exponentials with rate 1, Y_1 and Y_2
STEP 2: If $Y_2 \geq (Y_1 - 1)^2/2$, set $X = Y_1$. Otherwise return to Step 1.

Suppose now that the foregoing results in Y_1 being accepted—and so we know that Y_2 is larger than $(Y_1-1)^2/2$. By how much does the one exceed the other? To answer this, recall that Y_2 is exponential with rate 1, and so, given that it exceeds some value, the amount by which Y_2 exceeds $(Y_1-1)^2/2$ [i.e., its "additional life" beyond the time $(Y_1-1)^2/2$] is (by the memoryless property) also exponentially distributed with rate 1. That is, when we accept in Step 2 we obtain not only X (the absolute value of a unit normal) but by computing $Y_2 - (Y_1 - 1)^2/2$ we can also generate an exponential random variable (independent of X) having rate 1.

Hence, summing up, we have the following algorithm that generates an exponential with rate 1 and an independent unit normal random variable.

STEP 1: Generate Y_1, an exponential random variable with rate 1.
STEP 2: Generate Y_2, an exponential random variable with rate 1.
STEP 3: If $Y_2 - (Y_1 - 1)^2/2 > 0$, set $Y = Y_2 - (Y_1 - 1)^2/2$ and go to Step 4. Otherwise go to Step 1.
STEP 4: Generate a random number U and set

$$Z = \begin{cases} Y_1 & \text{if } U \le \frac{1}{2} \\ -Y_1 & \text{if } U > \frac{1}{2} \end{cases}$$

The random variables Z and Y generated by the foregoing are independent with Z being normal with mean 0 and variance 1 and Y being exponential with rate 1. (If you want the normal random variable to have mean μ and variance σ^2, just take $\mu + \sigma Z$.)

REMARKS

1. Since $c = \sqrt{2e/\pi} \approx 1.32$, the foregoing requires a geometric distributed number of iterations of Step 2 with mean 1.32.

2. If we want to generate a sequence of unit normal random variables, we can use the exponential random variable Y obtained in Step 3 as the initial exponential needed in Step 1 for the next normal to be generated. Hence, on the average, we can simulate a unit normal by generating 1.64 $(= 2 \times 1.32 - 1)$ exponentials and computing 1.32 squares. ■

5.3. The Polar Method for Generating Normal Random Variables

Let X and Y be independent unit normal random variables and let R and θ denote the polar coordinates of the vector (X, Y). That is (see Figure 5.2),

$$R^2 = X^2 + Y^2$$

$$\tan \theta = \frac{Y}{X}$$

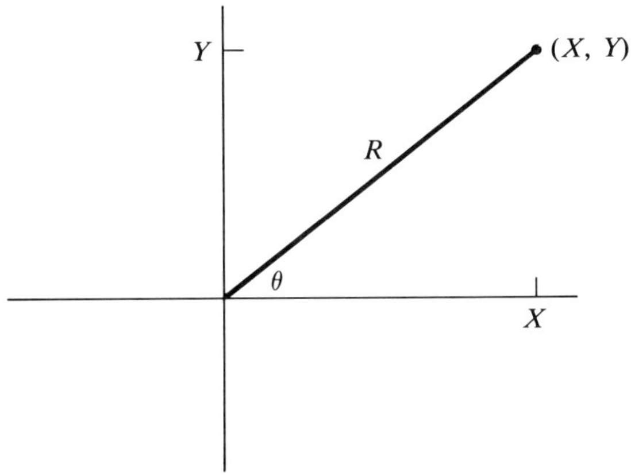

Figure 5.2

Since X and Y are independent, their joint density is the product of their individual densities and is thus given by

$$f(x, y) = \frac{1}{\sqrt{2\pi}} e^{-x^2/2} \frac{1}{\sqrt{2\pi}} e^{-y^2/2}$$

$$= \frac{1}{2\pi} e^{-(x^2+y^2)/2} \tag{5.3}$$

To determine the joint density of R^2 and Θ—call it $f_{R^2, \Theta}(d, \theta)$—we make the change of variables

$$d = x^2 + y^2, \qquad \theta = \tan^{-1}\left(\frac{y}{x}\right)$$

As the Jacobian of this transformation—that is, the determinant of partial derivatives of d and θ with respect to x and y—is easily shown to equal 2, it follows from Equation (5.3) that the joint density function of R^2 and Θ is given by

$$f_{R^2, \Theta}(d, \theta) = \frac{1}{2} \frac{1}{2\pi} e^{-d/2}, \qquad 0 < d < \infty, \quad 0 < \theta < 2\pi$$

However, as this is equal to the product of an exponential density having mean 2 (namely, $\frac{1}{2} e^{-d/2}$) and the uniform density on $(0, 2\pi)$ (namely, $1/2\pi$), it follows that

> R^2 and Θ are independent, with R^2 being exponential with mean 2 and Θ being uniformly distributed over $(0, 2\pi)$ (5.4)

We can now generate a pair of independent normal random variables X and Y by using (5.4) to first generate their polar coordinates and then transforming back to rectangular coordinates. This is accomplished as follows:

STEP 1: Generate random numbers U_1 and U_2.
STEP 2: $R^2 = -2 \log U_1$ (and thus R^2 is exponential with mean 2). Set $\Theta = 2\pi U_2$ (and thus Θ is uniform between 0 and 2π).
STEP 3: Now let

$$X = R \cos \Theta = \sqrt{-2 \log U_1} \cos(2\pi U_2)$$
$$Y = R \sin \Theta = \sqrt{-2 \log U_1} \sin(2\pi U_2) \tag{5.5}$$

The transformations given by Equations (5.5) are known as the Box–Muller transformations.

Unfortunately, the use of the Box–Muller transformations (5.5) to generate a pair of independent unit normals is computationally not very efficient: the reason for this being the need to compute the sine and cosine trigonometric functions. There is, however, fortuitously a way to get around this time-consuming difficulty by an indirect computation of the sine and cosine of a random angle (as opposed to a direct computation which generates U and then computes the sine and cosine of $2\pi U$). To begin, note that if U is uniform on (0, 1) then $2U$ is uniform on (0, 2) and so $2U - 1$ is uniform on $(-1, 1)$. Thus, if we generate random numbers U_1 and U_2 and set

$$V_1 = 2U_1 - 1$$
$$V_2 = 2U_2 - 1$$

then (V_1, V_2) is uniformly distributed in the square of area 4 centered at (0, 0)—see Figure 5.3.

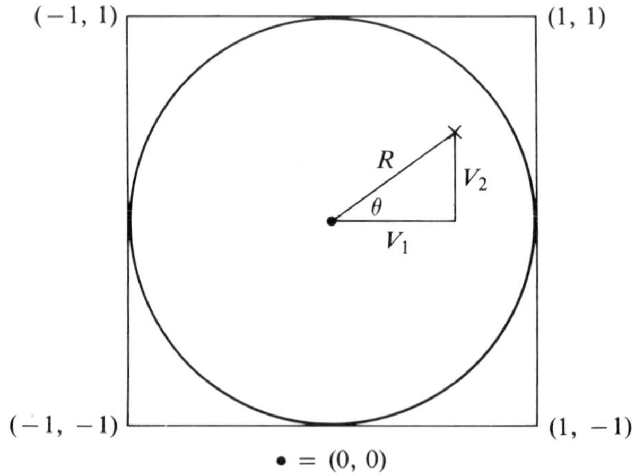

Figure 5.3

Suppose now that we continually generate such pairs (V_1, V_2) until we obtain one that is contained in the circle of radius 1 centered at (0, 0)—that is, until (V_1, V_2) is such that $V_1^2 + V_2^2 \leq 1$. It now follows that such a pair (V_1, V_2) is uniformly distributed in the circle. If we let R and Θ denote the polar coordinates of this pair then it is not difficult to verify that R and Θ are independent, with R^2 being uniformly distributed on (0, 1) (see Problem 18) and with Θ being uniformly distributed over $(0, 2\pi)$. Since Θ is thus a random angle, it follows that we can generate the sine and cosine of a random angle Θ by generating a random point (V_1, V_2) in the circle and then setting

$$\sin\Theta = \frac{V_2}{R} = \frac{V_2}{(V_1^2 + V_2^2)^{1/2}}$$

$$\cos\Theta = \frac{V_1}{R} = \frac{V_1}{(V_1^2 + V_2^2)^{1/2}}$$

It now follows from the Box–Muller transformation (5.5) that we can generate independent unit normals by generating a random number U and setting

$$X = (-2\log U)^{1/2}\frac{V_1}{(V_1^2 + V_2^2)^{1/2}}$$

$$Y = (-2\log U)^{1/2}\frac{V_2}{(V_1^2 + V_2^2)^{1/2}} \tag{5.6}$$

In fact, since $R^2 = V_1^2 + V_2^2$ is itself uniformly distributed over (0, 1) and is independent of the random angle Θ, we can use it as the random number U needed in Equations (5.6). Therefore, letting $S = R^2$, we obtain that

$$X = (-2\log S)^{1/2}\frac{V_1}{S^{1/2}} = V_1\left(\frac{-2\log S}{S}\right)^{1/2}$$

$$Y = (-2\log S)^{1/2}\frac{V_2}{S^{1/2}} = V_2\left(\frac{-2\log S}{S}\right)^{1/2}$$

are independent unit normals when (V_1, V_2) is a randomly chosen point in the circle of radius 1 centered at the origin, and $S = V_1^2 + V_2^2$.

Summing up we thus have the following approach to generating a pair of independent unit normals:

STEP 1: Generate random numbers U_1 and U_2.
STEP 2: Set $V_1 = 2U_1 - 1$, $V_2 = 2U_2 - 1$, $S = V_1^2 + V_2^2$.
STEP 3: If $S > 1$ return to Step 1.
STEP 4: Return the independent unit normals

$$X = \sqrt{\frac{-2\log S}{S}}\,V_1, \qquad Y = \sqrt{\frac{-2\log S}{S}}\,V_2$$

The above is called the polar method. Since the probability that a random point in the square will fall within the circle is equal to $\pi/4$ (the area of the circle divided by the area of the square), it follows that, on average, the polar method will require $4/\pi = 1.273$ iterations of Step 1. Hence it will, on average, require 2.546 random numbers, 1 logarithm, 1 square root, 1 division, and 4.546 multiplications to generate two independent unit normals.

5.4. Generating a Poisson Process

Suppose we wanted to generate the first n event times of a Poisson process with rate λ. To do so we make use of the result that the times between successive events for such a process are independent exponential random variables each with rate λ. Thus one way to generate the process is to generate these interarrival times. So if we generate n random numbers $U_1, U_2, \ldots, U_n$ and set $X_i = -1/\lambda \log U_i$, then X_i can be regarded as the time between the $(i-1)$st and the ith event of the Poisson process. Since the actual time of the jth event will equal the sum of the first j interarrival times, it thus follows that the generated values of the first n event times are $\sum_{i=1}^{j} X_i$, $j = 1, \ldots, n$.

If we wanted to generate the first T time units of the Poisson process, we can follow the above procedure of successively generating the interarrival times, stopping when their sum exceeds T. That is, the following algorithm can be used to generate all the event times occurring in $(0, T)$ of a Poisson process having rate λ. In the algorithm t refers to time, I is the number of events that have occurred by time t, and $S(I)$ is the most recent event time.

GENERATING THE FIRST T TIME UNITS OF A POISSON PROCESS WITH RATE λ

STEP 1: $t = 0$, $I = 0$.
STEP 2: Generate a random number U.
STEP 3: $t = t - \frac{1}{\lambda} \log U$. If $t > T$, stop.
STEP 4: $I = I + 1$, $S(I) = t$.
STEP 5: Go to Step 2.

The final value of I in the above algorithm will represent the number of events that occur by time T, and the values $S(1), \ldots, S(I)$ will be the I event times in increasing order.

There is another approach for simulating the first T time units of a Poisson process that is also quite efficient. It starts by simulating $N(T)$—the total number of events that occur by time T—and then makes use of a result which states that, given $N(T)$, the times at which these events occur are distributed independently and uniformly over $(0, T)$. [This is quite intuitive since by the stationary increment axiom it is intuitive that an arbitrary event time is uniformly distributed over the interval, and by the independent increment axiom it is intuitive that, given $N(T)$, these $N(T)$ event times would be independent.] Hence we can start by generating the value of $N(T)$, a Poisson random variable with mean λT (by the approach presented in Section 4.3 of Chapter 4 if λ is large or either by that approach or by the method outlined in Section 5.2 of this chapter if λ is not large). If the generated value of $N(T)$ is n, we then generate n random numbers—call them $U_1, \ldots, U_n$—and, as TU_i will be uniformly distributed over $(0, T)$, the set of event times will thus be $\{TU_1, \ldots, TU_n\}$. If we were to stop here, this approach would certainly be more efficient than simulating the exponentially distributed interarrival times. However, we usually desire the event times in increasing order [e.g., so as to be able to know $N(s)$ for all $s < T$]; and so we would also need to order the values TU_i, $i = 1, \ldots, n$.

5.5. Generating a Nonhomogeneous Poisson Process

An extremely important counting process for modeling purposes is the nonhomogeneous Poisson process, which relaxes the Poisson process assumption of stationary increments. Thus it allows for the possibility that the arrival rate need not be constant but can vary with time. It is usually very difficult to obtain analytical results for a mathematical model that assumes a nonhomogeneous Poisson arrival process, and as a result such processes are not applied as often as they should be. However, since the use of simulation helps one to analyze such models, we expect that such mathematical models will become more common.

Suppose that we wanted to simulate the first T time units of a nonhomogeneous Poisson process with intensity function $\lambda(t)$. The first method we present, called the *thinning* or *random sampling* approach, starts by choosing a value λ which is such that

$$\lambda(t) \leq \lambda \qquad \text{for all } t \leq T$$

Now as shown in Chapter 2, such a nonhomogeneous Poisson process can be generated by a random selection of the event times of a Poisson process having rate λ. That is, if an event of a Poisson process with rate λ that occurs at time t is counted (independently of what has transpired previously) with probability $\lambda(t)/\lambda$, then the process of counted events is a nonhomogeneous Poisson process with intensity function $\lambda(t)$, $0 \leq t \leq T$. Hence, by simulating a Poisson process and then randomly counting its events, we can generate the desired nonhomogeneous Poisson process. This can be written algorithmically as follows.

GENERATING THE FIRST T TIME UNITS OF A NONHOMOGENEOUS POISSON PROCESS

STEP 1: $t = 0$, $I = 0$.
STEP 2: Generate a random number U.

STEP 3: $t = t - \frac{1}{\lambda} \log U$. If $t > T$, stop.
STEP 4: Generate a random number U.
STEP 5: If $U \leq \lambda(t)/\lambda$, set $I = I + 1$, $S(I) = t$.
STEP 6: Go to Step 2.

In the above $\lambda(t)$ is the intensity function and λ is such that $\lambda(t) \leq \lambda$. The final value of I represents the number of events by time T, and $S(1), \ldots, S(I)$ are the event times.

The above procedure, referred to as the thinning algorithm (because it "thins" the homogeneous Poisson points), is clearly most efficient, in the sense of having the fewest number of rejected events times, when $\lambda(t)$ is near λ throughout the interval. Thus an obvious improvement is to break up the interval into subintervals and then use the procedure over each subinterval. That is, determine appropriate values k, $0 = t_0 < t_1 < t_2 < \cdots < t_k < t_{k+1} = T$, $\lambda_1, \ldots, \lambda_{k+1}$ such that

$$\lambda(s) \leq \lambda_i \qquad \text{if } t_{i-1} \leq s < t_i, \quad i = 1, \ldots, k+1 \tag{5.7}$$

Now generate the nonhomogeneous Poisson process over the interval (t_{i-1}, t_i) by generating exponential random variables with rate λ_i and accepting the generated event occurring at time s, $s \in (t_{i-1}, t_i)$, with probability $\lambda(s)/\lambda_i$. Because of the memoryless property of the exponential and the fact that the rate of an exponential can be changed upon multiplication by a constant, it follows that there is no loss of efficiency in going from one subinterval to the next. That is, if we are at $t \in (t_{i-1}, t_i)$ and generate X, an exponential with rate λ_i, which is such that $t + X > t_i$, then we can use $\lambda_i[X - (t_i - t)]/\lambda_{i+1}$ as the next exponential with rate λ_{i+1}.

We thus have the following algorithm for generating the first T time units of a nonhomogeneous Poisson process with intensity function $\lambda(s)$ when the relations (5.7) are satisfied. In the algorithm t represents the present time, J the present interval (i.e., $J = j$ when $t_{j-1} \leq t < t_j$), I the number of events so far, and $S(1), \ldots, S(I)$ the event times.

Generating the First T Time Units of a Nonhomogeneous Poisson Process

STEP 1: $t = 0$, $J = 1$, $I = 0$.
STEP 2: Generate a random number U and set $X = \frac{-1}{\lambda_J} \log U$.
STEP 3: If $t + X > t_J$, go to Step 8.
STEP 4: $t = t + X$.
STEP 5: Generate a random number U.
STEP 6: If $U \leq \lambda(t)/\lambda_J$, set $I = I + 1$, $S(I) = t$.
STEP 7: Go to Step 2.
STEP 8: If $J = k + 1$, stop.
STEP 9: $X = (X - t_J + t)\lambda_J/\lambda_{J+1}$, $t = t_J$, $J = J + 1$.
STEP 10: Go to Step 3.

Suppose now that over some subinterval (t_{i-1}, t_i) it follows that $\underline{\lambda}_i > 0$, where

$$\underline{\lambda}_i \equiv \text{Infimum } \{\lambda(s)\colon t_{i-1} \leq s < t_i\}$$

In such a situation we should not use the thinning algorithm directly but rather should first simulate a Poisson process with rate $\underline{\lambda}_i$ over the desired interval and then simulate a nonhomogeneous Poisson process with the intensity function $\lambda(s) = \lambda(s) - \underline{\lambda}_i$ when $s \in (t_{i-1}, t_i)$. (The final exponential generated for the Poisson process, which carries one beyond the desired boundary, need not be wasted but can be suitably transformed so as to be reusable.) The superposition (or merging) of the two processes yields the desired process over the interval. The reason for doing it this way is that it saves the need to generate uniform random variables for a Poisson distributed number, with mean $\underline{\lambda}_i(t_i - t_{i-1})$, of the event times. For example, consider the case where

$$\lambda(s) = 10 + s, \qquad 0 < s < 1$$

Using the thinning method with $\lambda = 11$ would generate an expected number of 11 events, each of which would require a random number to determine whether or not to accept it. On the other hand, to generate a Poisson process with rate 10 and then merge it with a nonhomogeneous Poisson process with rate $\lambda(s) = s$, $0 < s < 1$ (generated by the thinning algorithm with $\lambda = 1$), would yield an equally distributed number of event times but with the expected number needing to be checked to determine acceptance being equal to 1.

A second method for simulating a nonhomogeneous Poisson process having intensity function $\lambda(t)$, $t > 0$, is to directly generate the successive event times. So let $S_1, S_2, \ldots$ denote the successive event times of such a process. As these random variables are clearly dependent, we generate them in sequence—starting with S_1, and then using the generated value of S_1 to generate S_2, and so on.

To start, note that if an event occurs at time s, then, independent of what has occurred prior to s, the additional time until the next event has the distribution F_s, given by

$$\begin{aligned}
F_s(x) &= P\{\text{time from } s \text{ until next event is less than } x \mid \text{event at } s\} \\
&= P\{\text{next event is before } x + s \mid \text{event at } s\} \\
&= P\{\text{event between } s \text{ and } s + x \mid \text{event at } s\} \\
&= P\{\text{event between } s \text{ and } s + x\} \qquad \text{by independent increments} \\
&= 1 - P\{0 \text{ events in } (s, s + x)\} \\
&= 1 - \exp\left(-\int_s^{s+x} \lambda(y)\, dy\right) \\
&= 1 - \exp\left(-\int_0^x \lambda(s + y)\, dy\right) \qquad (5.8)
\end{aligned}$$

We can now simulate the event times $S_1, S_2, \ldots$ by generating S_1 from the distribution F_0; then if the simulated value of S_1 is s_1, we generate S_2 by adding s_1 to a generated value from the distribution F_{s_1}, and if this sum is s_2 we generate S_3 by adding s_2 to a generated value from the distribution F_{s_2} and so on. The method used to simulate from these distributions should of course depend on their form. In the following example the distributions F_s are easily inverted and so the inverse transform method can be applied.

Example 5g Suppose that $\lambda(t) = 1/(t + a)$, $t \geq 0$, for some positive constant a. Then

$$\int_0^x \lambda(s + y)\, dy = \int_0^x \frac{1}{s + y + a}\, dy = \log\left(\frac{x + s + a}{s + a}\right)$$

Hence, from Equation (5.8),

$$F_s(x) = 1 - \frac{s + a}{x + s + a} = \frac{x}{x + s + a}$$

To invert this, suppose that $x = F_s^{-1}(u)$, and so

$$u = F_s(x) = \frac{x}{x + s + a}$$

or, equivalently,

$$x = \frac{u(s + a)}{1 - u}$$

That is,

$$F_s^{-1}(u) = (s + a)\frac{u}{1 - u}$$

We can therefore generate the successive event times $S_1, S_2, \ldots$ by generating random numbers $U_1, U_2, \ldots$ and then recursively setting

$$S_1 = \frac{aU_1}{1 - U_1}$$

$$S_2 = S_1 + (S_1 + a)\frac{U_2}{1 - U_2} = \frac{S_1 + aU_2}{1 - U_2}$$

and, in general,

$$S_j = S_{j-1} + (S_{j-1} + a)\frac{U_j}{1 - U_j} = \frac{S_{j-1} + aU_j}{1 - U_j}, \qquad j \geq 2$$

■

Problems

1. Give a method for generating a random variable having density function

$$f(x) = e^x/(e - 1), \qquad 0 \le x \le 1$$

2. Give a method to generate a random variable having density function

$$f(x) = \begin{cases} \dfrac{(x-2)}{2} & \text{if } 2 \le x \le 3 \\ \dfrac{(2 - x/3)}{2} & \text{if } 3 \le x \le 6 \end{cases}$$

3. Use the inverse transform method to generate a random variable having distribution function

$$F(x) = \frac{x^2 + x}{2}, \qquad 0 \le x \le 1$$

4. Give a method for generating a random variable having distribution function

$$F(x) = 1 - exp(-\alpha x^{\beta}), \qquad 0 < x < \infty$$

A random variable having such a distribution is said to be a Weibull random variable.

5. Give a method for generating a random variable having density function

$$f(x) = \begin{cases} e^{2x}, & -\infty < x < 0 \\ e^{-2x}, & 0 < x < \infty \end{cases}$$

6. (The Composition Method). Suppose it is relatively easy to generate random variables from any of the distributions F_i, $i = 1, \ldots, n$. How could we generate a random variable having the distribution function

$$F(x) = \sum_{i=1}^{n} p_i F_i(x)$$

where p_i, $i = 1, \ldots, n$, are nonnegative numbers whose sum is 1?

7. Using the result of Problem 6, give algorithms for generating random variables from the following distributions.

(a) $F(x) = \dfrac{x + x^3 + x^5}{3}, \qquad 0 \le x \le 1$

(b) $F(x) = \begin{cases} \dfrac{1 - e^{-2x} + 2x}{3} & \text{if } 0 < x < 1 \\ \dfrac{3 - e^{-2x}}{3} & \text{if } 1 < x < \infty \end{cases}$

(c) $F(x) = \sum_{i=1}^{n} \alpha_i x^i, \qquad 0 \le x \le 1, \quad \text{where } \alpha_i \ge 0, \quad \sum_{i=1}^{n} \alpha_i = 1$

8. Give a method to generate a random variable having distribution function

$$F(x) = \int_0^1 x^y e^{-y}\, dy \qquad 0 \le x \le 1$$

Hint: Think in terms of the composition method of Problem 6. In particular, let F denote the distribution function of X, and suppose that the conditional distribution of X given that $Y = y$ is

$$P\{X \le x | Y = y\} = x^y, \qquad 0 \le x \le 1$$

9. Write an algorithm that can be used to generate exponential random variables in sets of 3. Compare the computational requirements of this method with the one presented after Example 5c which generates them in pairs.

10. Suppose it is easy to generate a random variable from any of the distributions F_i, $i = 1, \ldots, n$. How can we generate from the following distributions?

(a) $F(x) = \prod_{i=1}^{n} F_i(x)$

(b) $F(x) = 1 - \prod_{i=1}^{n} [1 - F_i(x)]$

[*Hint*: *If* X_i, $i = 1, \ldots, n$, are independent random variables, with X_i having distribution F_i, which random variable has distribution function F?]

11. Using the rejection method and the results of Problem 10, give two other methods, aside from the inverse transform method, that can be used to generate a random variable having distribution function

$$F(x) = x^n, \qquad 0 \le x \le 1$$

Discuss the efficiency of the three approaches to generating from F.

12. Let G be a distribution function with density g and suppose, for constants $a < b$, we want to generate a random variable from the distribution function

$$F(x) = \frac{G(x) - G(a)}{G(b) - G(a)}, \qquad a \leq x \leq b$$

(a) If X has distribution G, then F is the conditional distribution of X given what information?

(b) Show that the rejection method reduces in this case to generating a random variable X having distribution G and then accepting it if it lies between a and b.

13. Give two methods for generating a random variable having density function

$$f(x) = xe^{-x}, \qquad 0 \leq x < \infty$$

and compare their efficiency.

14. Give an algorithm that generates a random variable having density

$$f(x) = 30(x^2 - 2x^3 + x^4), \qquad 0 \leq x \leq 1$$

Discuss the efficiency of this approach.

15. In Example 5f we simulated a normal random variable by using the rejection technique with an exponential distribution with rate 1. Show that among all exponential density functions $g(x) = \lambda e^{-\lambda x}$ the number of iterations needed is minimized when $\lambda = 1$.

16. Write a program that generates normal random variables by the method of Example 5f.

17. Write a program that generates normal random variables by the polar method.

18. Let (X, Y) be uniformly distributed in a circle of radius 1. Show that if R is the distance from the center of the circle to (X, Y) then R^2 is uniform on (0, 1).

19. Write a program that generates the first T time units of a Poisson process having rate λ.

20. (a) Write a program that uses the thinning algorithm to generate the first 10 time units of a nonhomogeneous Poisson process with intensity function

$$\lambda(t) = 3 + \frac{4}{t + 1}$$

(b) Give a way to improve upon the thinning algorithm for this example.

21. Give an efficient algorithm to generate the first 10 time units of a nonhomogeneous Poisson process having intensity function

$$\lambda(t) = \begin{cases} \dfrac{t}{5}, & 0 < t < 5 \\ 1 + 5(t - 5), & 5 < t < 10 \end{cases}$$

CHAPTER 6

The Discrete Event Simulation Approach

Introduction

Simulating a probabilistic model involves generating the stochastic mechanisms of the model and then observing the resultant flow of the model over time. Depending on the reasons for the simulation, there will be certain quantities of interest that we will want to determine. However, because the model's evolution over time often involves a complex logical structure of its elements, it is not always apparent how to keep track of this evolution so as to determine these quantities of interest. A general framework, built around the idea of "discrete events," has been developed to help one follow a model over time and determine the relevant quantities of interest. The approach to simulation based on this framework is often referred to as the *discrete event simulation approach*.

6.1. Simulation via Discrete Events

The key elements in a discrete event simulation are variables and events. To do the simulation we continually keep track of certain variables. In general, there are three types of variables that are often utilized—the time variable, counter variables, and the system state variable.

VARIABLES

1. Time variable t	Refers to the amount of (simulated) time that has elapsed
2. Counter variables	These variables keep a count of the number of times that certain events have occurred by time t
3. System state variable	This describes the "state of the system" at the time t

Whenever an "event" occurs the values of the above variables are changed, or updated, and we collect, as output, any relevant data of interest. In order to determine when the next event will occur, an "event list," which lists the nearest future events and when they are scheduled to occur, is maintained. Whenever an event "occurs" we then reset the time and all state and counter variables and collect the relevant data. In this way we are able to "follow" the system as it evolves over time.

As the above is only meant to give a rough idea of the elements of a discrete event simulation, it is useful to look at some examples. In Sections 6.2 and 6.3 we consider simulations of a waiting line, or queueing, system and a machine repair model.

6.2. A Single Server Queueing System

Consider a service system in which customers arrive according to the following stochastic mechanism: the time at which the first customer arrives has distribution function F_0; thereafter, if a customer arrives at time s then the time until the next arrival has distribution function F_s. There is a single server, and upon arrival a customer either enters service if this server is free at that moment or else joins the waiting queue if the server is busy. When the server completes serving a customer it then either begins serving the customer that had been waiting the longest (the so-called "first come first served" discipline) if there are any waiting customers, or, if there are no waiting customers, it remains free until the next customer's arrival. The amount of time it takes to service a customer is a random variable (independent of all other service times and of the arrival process), having probability distribution G. In addition, there is a fixed time T after which no additional arrivals are allowed to enter the system, although the server completes servicing all those that are already in the system at time T.

Suppose that we are interested in simulating the above system to determine such quantities as (a) the average time a customer spends in the system and (b) the average time past T that the last customer departs—that is, the average time at which the server can go home.

To do a simulation of the above system we use the following variables:

1. Time variable t
2. Counter variables N_A: the number of arrivals (by time t)
   N_D: the number of departures (by time t)
3. System state variable n: the number of customers in the system (at time t)

Since the natural time to change the above quantities is when there is either an arrival or a departure, we take these as the "events." That is, there are two types of event: arrivals and departures. The event list contains the times of the next arrival and departure (starting from time t). That is, the event list usually looks like

$$\text{Event List: } t_A,\ t_D$$

where t_A and t_D are, respectively, the times of the next arrival and departure.

We make use of the output variables $A(i)$ and $D(i)$, which are, respectively, the arrival and departure time of the ith customer. Also, we define the output variable T_p to equal the time past T that the last customer departs (and so T_p is 0 if there are no customers in the system at time T).

We are now ready to show how the above model can be simulated. To begin, we initialize the variables and the event list as follows:

INITIALIZE

Generate a random variable X having distribution F_0 (X is the first arrival time).

If $X > T$, end this simulation run with the output data: $N_A = 0$, $T_p = 0$.

If $X \leq T$, set $t = n = N_A = N_D = 0$. Set Event List: $t_A = X$.

The system is always updated by moving along the time axis until the next event occurs. To see how this is accomplished we need to consider cases.

Case 1 The only event on the event list is an arrival at time t_A (this will be the situation when the state variable n is 0).

Reset: $t = t_A$ (i.e., we move along to time t_A)
Reset: $n = n + 1$ (because now at time t_A there is one more customer)
Reset: $N_A = N_A + 1$ (because of the additional arrival at t_A)
Collect: The output data $A(N_A) = t$ (because customer number N_A arrived at the reset time t)
Generate: Random variables X, having distribution F_t (representing the time until the next arrival), and S, having distribution G (representing the next service time)
Reset: If $t + X > T$, then Event List: $t_D = t + S$.
If $t + X < T$, then Event List: $t_A = t + X$, $t_D = t + S$.
(Note that since $t + X$ is the time of the next arrival, we do not include it on the event list if it occurs after time T.)

Case 2 Event List: t_A, t_D, where $t_A \leq t_D$

Reset: $t = t_A$
Reset: $n = n + 1$
Reset: $N_A = N_A + 1$
Collect: The output data $A(N_A) = t$

Generate: X having distribution F_t
Reset: If $t + X > T$, then Event List: t_D
If $t + X \leq T$, then Event List: $t_A = t + X$, t_D
(Note that the next departure time remains unchanged.)

Case 3 Event List: t_A, t_D, where $t_D < t_A$
Reset: $t = t_D$
Reset: $n = n - 1$
Reset: $N_D = N_D + 1$ (since a departure occurs at time t_D)
Collect: The output data $D(N_D) = t$ (since customer N_D departed at t_D)
If $n > 0$: Generate a random variable Y from G: Reset $t_D = t + Y$
Reset Event List: t_A, t_D
If $n = 0$: Reset Event List: t_A

Case 4 Event List: t_D
Do exactly as in Case 3 except that if the reset value of n is $n = 0$, then reset Event List: $\varnothing$ (i.e., there are no events on the event list).

Case 5 Event List: $\varnothing$
In the case of an empty event list, stop the simulation and collect the data $T_p = \text{Max}(0, t - T)$. ∎

The above is illustrated in the flow diagram presented in Figure 6.1. Each time we pass through Case 5 we would have collected the data N_A, the total number of arrivals, which will equal N_D, the total number of departures. For each i, $i = 1, \ldots, N_A$, we have $A(i)$ and $D(i)$, the respective arrival and departure times of customer i [and thus $D(i) - A(i)$ represents the amount of time that customer i spent in the system]. Finally, we will have T_p, the time past T at which the last customer departed. Each time we collect the above data we say that a simulation run has been completed. After each run we then reinitialize and generate another run until it has been decided that enough data have been collected. (In Chapter 7 we consider the question of when to end the simulation.) The average of all the values of T_p that have been generated will be our estimate of the mean time past T that the last customer departs; similarly, the average of all the observed values of $D - A$ (i.e., the average time, over all customers observed in all our simulation runs, that a customer spends in the system) will be our estimate of the average time that a customer spends in the system.

6.3. A Repair Problem

A system needs n working machines to be operational. To guard against machine breakdown, additional machines are kept available as spares. Whenever a machine breaks down it is immediately replaced by a spare and

Initialize: $t = n = N_A = N_D = 0$. Generate $X \sim F_t$, and set $t_A = X$.
If $t_A > T$, end this simulation run.
If $t_A \leq T$, then Event List: t_A

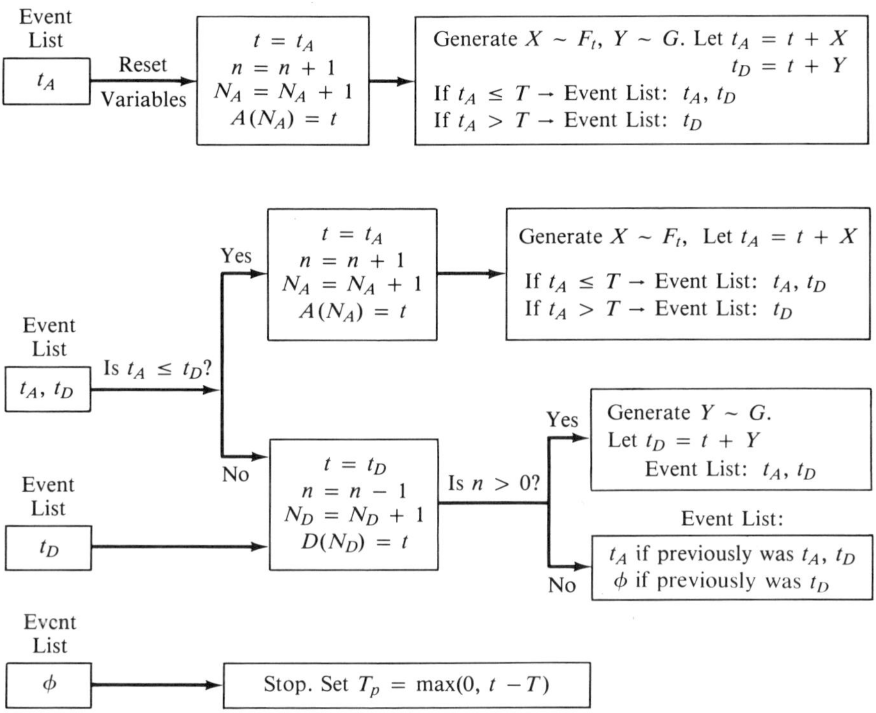

Figure 6.1

is itself sent to the repair facility, which consists of a single repairperson who repairs failed machines one at a time. Once a failed machine has been repaired it becomes available as a spare to be used when the need arises (see Figure 6.2). All repair times are independent random variables having the common distribution function G. Each time a machine is put into use the amount of time it functions before breaking down is a random variable, independent of the past, having distribution function F.

The system is said to "crash" when a machine fails and no spares are available. Assuming that there are initially $n + s$ functional machines of which n are put in use and s are kept as spares, we are interested in simulating this system so as to approximate $E[T]$, where T is the time at which the system crashes.

To simulate the above we utilize the following variables.

1. Time variable $\quad t$
2. System state variable $\quad$ r: the number of machines that are down at time t

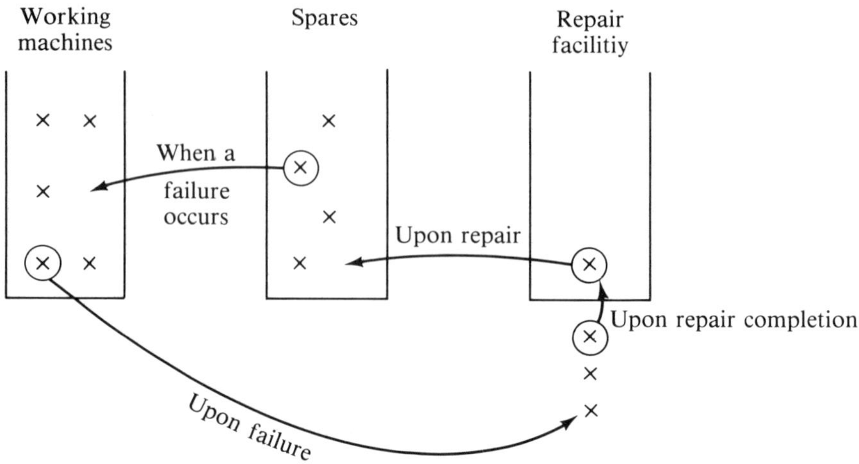

Figure 6.2 Repair model.

Since the system state variable will change either when a working machine breaks down or when a repair is completed, we say that an "event" occurs whenever either of these occurs. In order to know when the next event will occur, we need to keep track of the times at which the machines presently in use will fail and the time at which the machine presently being repaired (if there is a machine in repair) will complete its repair. Because we will always need to determine the smallest of the n failure times, it is convenient to store these n times in an ordered list. Thus it is convenient to let the event list be as follows:

$$\text{Event List: } t_1 \le t_2 \le t_3 \le \cdots \le t_n,\ t_{\text{rep}}$$

where $t_1, \ldots, t_n$ are the times (in order) at which the n machines presently in use will fail; and t_{rep}, which will only appear on the event list when there is a machine being repaired, is the time at which the machine presently being repaired will become operational. (Note that t_{rep} will only appear on the event list when the state variable r is greater than 0.)

To begin the simulation we initialize the above quantities as follows:

INITIALIZE

Set $t = r = 0$.

Generate $X_1, \ldots X_n$ independent random variables, each having distribution function F. Order the values $X_1, \ldots, X_n$ and let t_i be the ith smallest value, $i = 1, \ldots, n$.

Set Event List: $t_1, \ldots, t_n$.

Updating of the system proceeds according to the following two cases.

Case 1 Either $t_1 < t_{rep}$ or t_{rep} does not appear on the event list (because $r = 0$).
Reset: $t = t_1$
Reset: $r = r + 1$ (because another machine has failed)

If $r = s + 1$, stop this run and collect the data $T = t$ (since, as there are now $s + 1$ machines down, no spares are available).

If $r < s + 1$, generate a random variable X having distribution F. This random variable will represent the working time of the spare, which will now be put into use. Now reorder the values $t_2, t_3, \ldots, t_n, t + X$ and let t_i be the ith smallest of these values, $i = 1, \ldots, n$. That is (with $t_{n+1} = \infty$), if

$$t_j \leq t + X \leq t_{j+1} \qquad \text{reset} \qquad \begin{aligned} t_i &= t_{i+1}, \quad i = 1, \ldots, j-1 \\ t_j &= t + X \\ t_i &= t_i, \quad i = j+1, \ldots, n \end{aligned}$$

If $r = 1$, generate a random variable Y having distribution function G and put $t_{rep} = t + Y$ on the event list. (This is necessary because in this case the machine that has just failed is the only failed machine and thus repair will immediately begin on it; Y will be its repair time and so its repair will be completed at time $t + Y$.)

Case 2 Event List: $t_1, \ldots, t_n, t_{rep}$ and $t_{rep} \leq t_1$
Reset: $t = t_{rep}$
Reset: $r = r - 1$

If $r = 0$, eliminate t_{rep} from the event list.

If $r > 0$, generate a random variable Y having distribution function G, and representing the repair time of the machine just entering service, and change the value of t_{rep} on the event list to $t_{rep} = t + Y$. ■

The above rules for updating are illustrated in Figure 6.3.

Each time we stop (which occurs when $r = s + 1$) we say that a run is completed. The output for the run is the value of the crash time T. We then reinitialize and simulate another run. In all, we do a total of, say, k runs with the successive output variables being $T_1, \ldots, T_k$. Since these k random variables are independent and each represents a crash time, their average—that is, $\sum_{i=1}^{k} T_i/k$, is the estimate of $E[T]$, the mean crash time. The question of determining when to stop the simulation—that is, determining the value of k—is considered in Chapter 7, which presents the methods to statistically analyze the output from simulation runs.

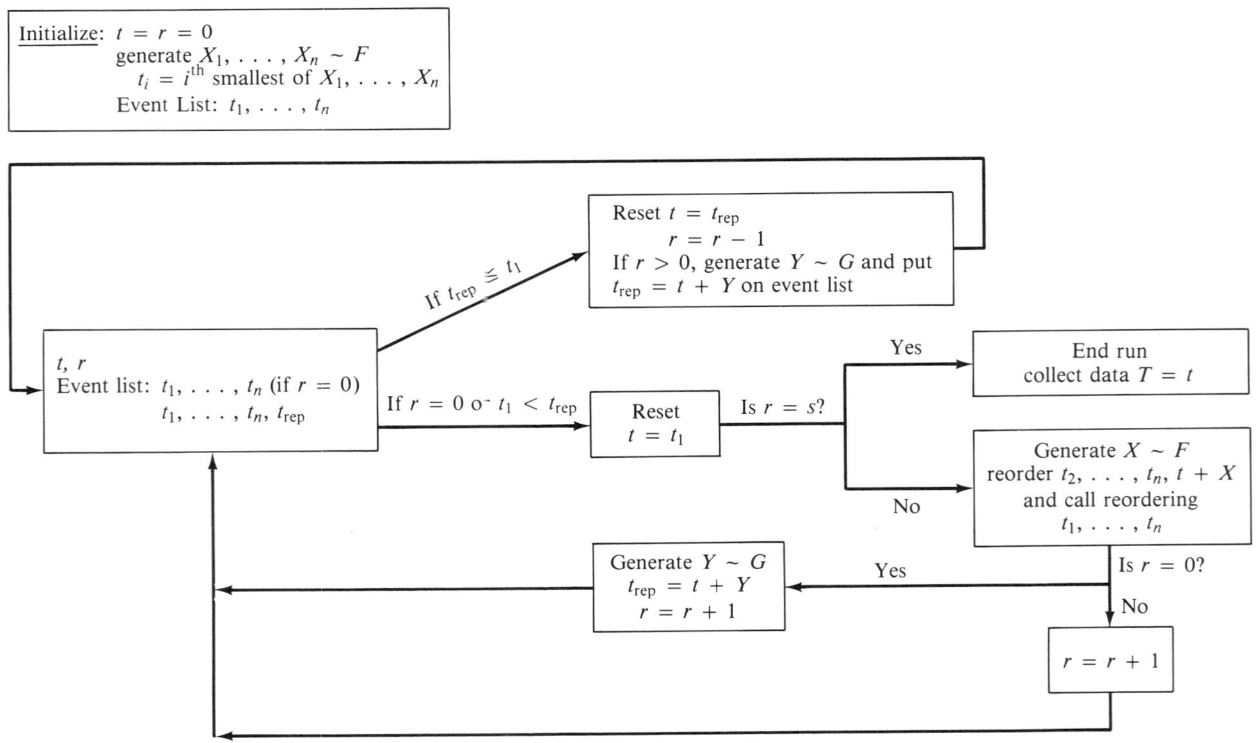
Initialize: t = r = 0
generate X1, . . . , Xn ~ F
ti = ith smallest of X1, . . . , Xn
Event List: t1, . . . , tn
t, r
Event list: t1, . . . , tn (if r = 0)
t1, . . . , tn, trep
If trep ≦ t1
Reset t = trep
r = r − 1
If r > 0, generate Y ~ G and put
trep = t + Y on event list
If r = 0 o- t1 < trep
Reset
t = t1
Is r = s?
Yes
No
End run
collect data T = t
Generate X ~ F
reorder t2, . . . , tn, t + X
and call reordering
t1, . . . , tn
Is r = 0?
Yes
No
Generate Y ~ G
trep = t + Y
r = r + 1
r = r + 1

Figure 6.3

6.4. Verification of the Simulation Model

The end product of the discrete event approach to simulation is a computer program which is hopefully free of error. To verify that there are indeed no bugs in the program, one should, of course, use all the "standard" techniques of debugging computer programs. However, there are several techniques that are particularly applicable in debugging simulation models, and we now discuss some of them.

As with all large programs one should attempt to debug in "modules" or subroutines. That is, one should attempt to break down the program into small and manageable entities that are logical wholes and then attempt to debug these entities. For example, in simulation models the generation of random variables constitutes one such module, and these modules should be checked separately.

The simulation should always be written broadly with a large number of input variables. Oftentimes by choosing suitable values we can reduce the simulation model to one that can be evaluated analytically or that has been previously extensively studied, so as to compare our simulated results with known answers.

In the testing stage, the program should be written to give as output all the random quantities it generates. By suitably choosing simple special cases, we can then compare the simulated output with the answer worked out by hand (and verified by others). For example, suppose we are simulating the first T time units of a k server queueing system. After inputting the values $T = 8$ (meant to be a small number) and $k = 2$, suppose the simulation program generates the following data:

Customer number:	1	2	3	4	5	6
Arrival time:	1.5	3.6	3.9	5.2	6.4	7.7
Service time:	3.4	2.2	5.1	2.4	3.3	6.2

and suppose that the program gives as output that the average time spent in the system by these six customers is 5.12.

However, by going through the calculations by hand (and having them verified by others), we see that the first customer spent 3.4 time units in the system; the second spent 2.2 (recall there are two servers); the third arrived at time 3.9, entered service at time 4.9 (when the first customer left), and spent 5.1 time units in service—thus customer 3 spent a time 6.1 in the system; customer 4 arrived at time 5.2, entered service at time 5.8 (when number 2 departed), and departed after an additional time 2.4—thus customer 4 spent a time 3.0 in the system; and so on. These calculations are presented below:

Arrival time:	1.5	3.6	3.9	5.2	6.4	7.7
Time when service began:	1.5	3.6	4.9	5.8	8.2	10.0
Departure time:	4.9	5.8	10.0	8.2	11.5	16.2
Time in system:	3.4	2.2	6.1	3.0	5.1	8.5

Hence the output for the average time spent in the system by all arrivals up to time $T = 8$ should have been

$$\frac{(3.4 + 2.2 + 6.1 + 3.0 + 5.1 + 8.5)}{6} = 4.71666 \cdots$$

thus showing that there is an error in the computer program which gave the output value 5.12.

A useful technique when searching for errors in the computer program is to utilize a *trace*. In a trace, the state variable, the event list, and the counter variables are all printed out after each event occurs. This allows one to follow the simulated system over time so as to determine when it is not performing as intended. (If no errors are apparent when following such a trace, one should then check the calculations relating to the output variables.)

Problems

1. Write a program to generate the desired output for the model of Section 6.2. Assume that

$$F_t(x) = 1 - e^{-\alpha x}, \qquad g(x) = \frac{\lambda e^{-\lambda x}(\lambda x)^{r-1}}{(r-1)!}, \quad x > 0$$

where g is the density function of a service time random variable. The quantities α, λ, r, and T should be input variables to the program. Use your program to estimate the desired quantities of interest when $\alpha = 10$, $\lambda = 50$, $r = 3$, $T = 9$. First try 100 runs and then 1000.

2. Suppose in the model of Section 6.2 we also wanted to obtain information about the amount of idle time a server would experience in a day. Explain how this could be accomplished.

3. Consider the model of Section 6.2 but suppose now that there are two servers, both of which have the same service distribution G, and that a customer can be served by either one. Write a flow diagram that can be used to simulate this model.

4. For the repair model presented in Section 6.3

(a) Write a computer program for this model.

(b) Use your program to estimate the mean crash time in the case where $n = 4$, $s = 3$, $F(x) = 1 - e^{-x}$, and $G(x) = 1 - e^{-2x}$.

5. In the model of Section 6.3, suppose that the repair facility consists of two servers, each of whom takes a random amount of time having distribution G to service a failed machine. Draw a flow diagram for this system.

6. A system experiences shocks that occur in accordance with a Poisson process having a rate of 1/hour. Each shock has a certain amount of damage associated with it. These damages are assumed to be independent random variables (which are also independent of the times at which the shocks occur), having the common density function

$$f(x) = xe^{-x}, \qquad x > 0$$

Damages dissipate in time at an exponential rate α—that is, a shock whose initial damage is x will have remaining damage value $xe^{-\alpha s}$ at time s after it occurs. In addition, the damage values are cumulative. (Thus, for example, if by time t there have been a total of two shocks, which originated at times t_1 and t_2 and had initial damages x_1 and x_2, then the total damage at time t is $\Sigma_{i=1}^{2} x_i e^{-\alpha(t-t_i)}$.) The system fails when the total damage exceeds some fixed constant C.

(a) Suppose we are interested in utilizing a simulation study to estimate the mean time that the system fails. Define the "events" and "variables" of this model and draw a flow diagram indicating how the simulation is to be run.
(b) Write a program that would generate k runs.
(c) Verify your program by comparing output with a by-hand calculation.
(d) With $\alpha = 0.5$, $C = 5$, and $k = 1000$, run your program and use the output to estimate the expected time until the system fails.

7. Messages arrive at a communications facility in accordance with a Poisson process having a rate of 2/hour. The facility consists of three channels and an arriving message will either go to a free channel if any of them are free or else will be lost if all channels are busy. The amount of time that a message ties up a channel is a random variable that depends on the weather condition at the time the message arrives. Specifically, if the message arrives when the condition is "good," then its processing time is a random variable having distribution function

$$F(x) = x, \qquad 0 < x < 1,$$

whereas if the condition is "bad" when a message arrives, then its processing time has distribution function

$$F(x) = x^3, \qquad 0 < x < 1$$

Initially, the condition is good, and it alternates between good and bad periods—with the good periods having fixed lengths of 2 hours and the bad periods having fixed lengths of 1 hour. (Thus, for example, at time 5 the condition changes from good to bad.)

Suppose we are interested in the distribution of the number of lost messages by time $T = 100$.

(a) Define the events and variables that enable us to use the discrete event approach.
(b) Write a flow diagram of the above.
(c) Write a program for the above.
(d) Verify your program by comparing an output with a hand calculation.
(e) Run your program to estimate the mean number of lost messages in the first 100 hours of operation.

C H A P T E R 7

Statistical Analysis of Simulated Data

Introduction

A simulation study is usually undertaken to determine the value of some quantity θ connected with a particular stochastic model. A simulation of the relevant system results in the output data X, a random variable whose expected value is the quantity of interest θ. A second independent simulation—that is, a second simulation run—provides a new and independent random variable having mean θ. This continues until we have amassed a total of k runs—and the k independent random variables $X_1, \cdots, X_k$—all of which are identically distributed with mean θ. The average of these k values, $\overline{X} = \Sigma_{i=1}^{k} X_i/k$, is then used as an estimator, or approximator, of θ.

In this chapter we consider the problem of deciding when to stop the simulation study—that is, deciding on the appropriate value of k. To help us decide when to stop, we will find it useful to consider the quality of our estimator of θ. In addition, we will also show how to obtain an interval in which we can assert that θ lies, with a certain degree of confidence.

The final two sections of this chapter show how we can estimate the quality of more complicated estimators than the sample mean—by using an important statistical technique known as "bootstrap estimators." The application of the bootstrap approach to simulation studies is given in the final section of this chapter.

7.1. The Sample Mean and Sample Variance

Suppose that $X_1, \cdots, X_n$ are independent random variables having the same distribution function. Let θ and σ^2 denote, respectively, their mean and variance—that is, $\theta = E[X_i]$ and $\sigma^2 = \text{Var}(X_i)$. The quantity

$$\overline{X} \equiv \sum_{i=1}^{n} \frac{X_i}{n}$$

which is the arithmetic average of the n data values, is called the *sample mean*. When the population mean θ is unknown, the sample mean is often used to estimate it.

Because

$$E[\overline{X}] = E\left[\sum_{i=1}^{n} \frac{X_i}{n}\right]$$

$$= \sum_{i=1}^{n} \frac{E[X_i]}{n}$$

$$= \frac{n\theta}{n} = \theta \tag{7.1}$$

it follows that $\overline{X}$ is an unbiased estimator of θ, where we say that an estimator of a parameter is an unbiased estimator of that parameter if its expected value is equal to the parameter.

To determine the "worth" of $\overline{X}$ as an estimator of the population mean θ, we consider its mean square error—that is, the expected value of the squared difference between $\overline{X}$ and θ. Now

$$E[(\overline{X} - \theta)^2] = \text{Var}(\overline{X}) \qquad (\text{since } E[\overline{X}] = \theta)$$

$$= \text{Var}\left(\frac{1}{n} \sum_{1}^{n} X_i\right)$$

$$= \frac{1}{n^2} \sum_{i=1}^{n} \text{Var}(X_i) \qquad (\text{by independence})$$

$$= \frac{\sigma^2}{n} \qquad (\text{since Var}(X_i) = \sigma^2) \tag{7.2}$$

Thus $\overline{X}$, the sample mean of the n data values $X_1, \cdots, X_n$, is a random variable with mean θ and variance σ^2/n. Since a random variable is unlikely to be too many standard deviations—equal to the square root of its variance—from its mean, it follows that $\overline{X}$ is a good estimator of θ when $\sigma/\sqrt{n}$ is small.

REMARK

The justification for the above statement that a random variable is unlikely to be too many standard deviations away from its mean follows from both the

Chebyshev inequality and, more importantly for simulation studies, from the central limit theorem. Indeed, for any $c > 0$, Chebyshev's inequality (see Section 2.7 of Chapter 2) yields the rather conservative bound

$$P\left\{|\overline{X} - \theta| > \frac{c\sigma}{\sqrt{n}}\right\} \leq \frac{1}{c^2}$$

However, when n is large, as will usually be the case in simulations, we can apply the central limit theorem to assert that $(\overline{X} - \theta)/(\sigma/\sqrt{n})$ is approximately distributed as a unit normal random variable; and thus

$$\begin{aligned} P\{|\overline{X} - \theta| > c\sigma/\sqrt{n}\} &\approx P\{|Z| > c\}, \qquad \text{where } Z \text{ is a unit normal} \\ &= 2[1 - \phi(c)] \end{aligned} \tag{7.3}$$

where ϕ is the unit normal distribution function. For example, since $\phi(1.96) = .95$, Equation (7.3) states that the probability that the sample mean differs by θ by more than $1.96\sigma/\sqrt{n}$ is approximately .05; whereas the weaker Chebyshev inequality only yields that this probability is less than $1/(1.96)^2 = .2603$. ■

The difficulty with directly using the value of σ^2/n as an indication of how well the sample mean of n data values estimates the population mean is that the population variance σ^2 is not usually known. Thus we also need to estimate it. Since

$$\sigma^2 = E[(X - \theta)^2]$$

is the average of the square of the difference between a datum value and its (unknown) mean, it might seem upon using $\overline{X}$ as the estimator of the mean that a natural estimator of σ^2 would be $\sum_{i=1}^{n}(X_i - \overline{X})^2/n$, the average of the squared distances between the data values and the estimated mean. However, to make the estimator unbiased (and for other technical reasons) we prefer to divide the sum of squares by $n - 1$ rather than n.

Definition:

The quantity S^2, defined by

$$S^2 = \frac{\sum_{i=1}^{n}(X_i - \overline{X})^2}{n - 1}$$

is called the sample variance.

Using the algebraic identity

$$\sum_{i=1}^{n}(X_i - \overline{X})^2 = \sum_{i=1}^{n} X_i^2 - n\overline{X}^2 \tag{7.4}$$

whose proof is left as an exercise, we now show that the sample variance is an unbiased estimator of σ^2.

Proposition:

$$E[S^2] = \sigma^2$$

Proof. Using the identity (7.4) we see that

$$\begin{aligned}(n-1)E[S^2] &= E\left[\sum_{i=1}^{n} X_i^2\right] - nE[\overline{X}^2] \\ &= nE[X_1^2] - nE[\overline{X}^2] \end{aligned} \tag{7.5}$$

where the last equality follows since the X_i all have the same distribution. Recalling that for any random variable Y, $\text{Var}(Y) = E[Y^2] - (E[Y])^2$ or, equivalently,

$$E[Y^2] = \text{Var}(Y) + (E[Y])^2$$

we obtain that

$$\begin{aligned} E[X_1^2] &= \text{Var}(X_1) + (E[X_1])^2 \\ &= \sigma^2 + \theta^2 \end{aligned}$$

and

$$\begin{aligned} E[\overline{X}^2] &= \text{Var}(\overline{X}) + (E[\overline{X}])^2 \\ &= \frac{\sigma^2}{n} + \theta^2 \qquad \text{[from (7.2) and (7.1)]} \end{aligned}$$

Thus, from Equation (7.5), we obtain that

$$(n-1)E[S^2] = n(\sigma^2 + \theta^2) - n\left(\frac{\sigma^2}{n} + \theta^2\right) = (n-1)\sigma^2$$

which proves the result. □

We use the sample variance S^2 as our estimator of the population variance σ^2; and we use $S = \sqrt{S^2}$, the so-called sample standard deviation, as our estimator of σ.

Suppose now that, as in a simulation, we have the option of continually generating additional data values X_i. If our objective is to estimate the value of $\theta = E[X_i]$, when should we stop generating new data values? The answer to this question is that we should first choose an acceptable value d for the standard deviation of our estimator—for if d is the standard deviation of the estimator $\overline{X}$, then we can, for example, be 95% certain that $\overline{X}$ will not differ from θ by more than $1.96d$. We should then continue to generate new data until we have generated n data values for which our estimate of $\sigma/\sqrt{n}$—namely, $S/\sqrt{n}$—is less than the acceptable value d. Since the sample standard deviation S may not be a particularly good estimate of σ (nor may the normal approximation be valid) when the sample size is small, we thus recommend the following procedure to determine when to stop generating new data values.

A Method for Determining When to Stop Generating New Data

1. Choose an acceptable value d for the standard deviation of the estimator.
2. Generate at least 30 data values.
3. Continue to generate additional data values, stopping when you have generated k values and $S/\sqrt{k} < d$, where S is the sample standard deviation based on those k values.
4. The estimate of θ is given by $\overline{X} = \sum_{i=1}^{k} X_i/k$.

Example 7a Consider a service system in which no new customers are allowed to enter after 5 P.M. Suppose that each day follows the same probability law and that we are interested in estimating the expected time at which the last customer departs the system. Furthermore, suppose we want to be at least 95% certain that our estimated answer will not differ from the true value by more than 15 seconds.

To satisfy the above requirement it is necessary that we continually generate data values relating to the time at which the last customer departs (each time by doing a simulation run) until we have generated a total of k values, where k is at least 30 and is such that $1.96S/\sqrt{k} < 15$—where S is the sample standard deviation (measured in seconds) of these k data values. Our estimate of the expected time at which the last customer departs will be the average of the k data values. ■

In order to use the above technique for determining when to stop generating new values, it would be valuable if we had a method for recursively computing the successive sample means and sample variances, rather than having to recompute from scratch each time a new datum value is generated. We now show how this can be done. Consider the sequence of data values $X_1, X_2, \cdots$, and let

$$\overline{X}_j = \sum_{i=1}^{j} \frac{X_i}{j}$$

and

$$S_j^2 = \sum_{i=1}^{j} \frac{(X_i - \overline{X}_j)^2}{j-1}, \qquad j \geq 2$$

denote, respectively, the sample mean and sample variance of the first j data values. The following recursion should be used to successively compute the current value of the sample mean and sample variance.

With $S_1^2 = 0$, $\overline{X}_0 = 0$.

$$\overline{X}_{j+1} = \overline{X}_j + \frac{X_{j+1} - \overline{X}_j}{j+1} \tag{7.6}$$

$$S_{j+1}^2 = \left(1 - \frac{1}{j}\right)S_j^2 + (j+1)(\overline{X}_{j+1} - \overline{X}_j)^2 \tag{7.7}$$

Example 7b If the first three data values are $X_1 = 5$, $X_2 = 14$, and $X_3 = 9$, then Equations (7.6) and (7.7) yield that

$$\begin{aligned}
\overline{X}_1 &= 5 \\
\overline{X}_2 &= 5 + \tfrac{9}{2} = \tfrac{19}{2} \\
S_2^2 &= 2(\tfrac{19}{2} - 5)^2 = \tfrac{81}{2} \\
\overline{X}_3 &= \tfrac{19}{2} + \tfrac{1}{3}(9 - \tfrac{19}{2}) = \tfrac{28}{3} \\
S_3^2 &= \tfrac{81}{4} + 3(\tfrac{28}{3} - \tfrac{19}{2})^2 = \tfrac{61}{3}
\end{aligned}$$

■

The analysis is somewhat modified when the data values are Bernoulli (or 0–1) random variables, as is the case when we are estimating a probability. That is, suppose we can generate random variables X_i such that

$$X_i = \begin{cases} 1 & \text{with probability } p \\ 0 & \text{with probability } 1-p \end{cases}$$

and suppose we are interested in estimating $E[X_i] = p$. Since, in this situation,

$$\text{Var}(X_i) = p(1-p)$$

there is no need to utilize the sample variance to estimate $\text{Var}(X_i)$. Indeed, if we have generated n values $X_1, \cdots, X_n$, then as the estimate of p will be

$$\overline{X}_n = \sum_{i=1}^{n} \frac{X_i}{n}$$

a natural estimate of $\text{Var}(X_i)$ is $\overline{X}_n(1 - \overline{X}_n)$. Hence, in this case, we have the following method for deciding when to stop.

1. Choose an acceptable value d for the standard deviation of the estimator.
2. Generate at least 30 data values.
3. Continue to generate additional data values, stopping when you have generated k values and $[\overline{X}_k(1 - \overline{X}_k)/k]^{1/2} < d$.
4. The estimate of p is $\overline{X}_k$, the average of the k data values.

Example 7c Suppose, in Example 7a, we were interested in estimating the probability that there was still a customer in the store at 5:30. To do so, we would simulate successive days and let

$$X_i = \begin{cases} 1 & \text{if there is a customer present at 5:30 on day } i \\ 0 & \text{otherwise} \end{cases}$$

We would simulate at least 30 days and continue to simulate until the kth day, where k is such that $[p_k(1 - p_k)/k]^{1/2} < d$, where $p_k = \overline{X}_k$ is the proportion of these k days in which there is a customer present at 5:30 and where d is an acceptable value for the standard deviation of the estimator p_k. ■

7.2. Interval Estimates of a Population Mean

Suppose again that $X_1, X_2, \cdots, X_n$ are independent random variables from a common distribution having mean θ and variance σ^2. Whereas the sample mean $\overline{X} = \sum_{i=1}^{n} X_i/n$ is an effective estimator of θ, we do not really expect that $\overline{X}$ will be equal to θ but rather that it will be "close." As a result, it is sometimes more valuable to be able to specify an interval for which we have a certain degree of confidence that θ lies within.

To obtain such an interval we need the (approximate) distribution of the estimator $\overline{X}$. To determine this, first recall, from Equations (7.1) and (7.2), that

$$E[\overline{X}] = \theta, \qquad \text{Var}(\overline{X}) = \frac{\sigma^2}{n}$$

and thus, from the central limit theorem, it follows that for large n

$$\sqrt{n}\,\frac{(\overline{X}-\theta)}{\sigma} \mathrel{\dot\sim} N(0,\ 1)$$

where $\dot\sim N(0, 1)$ means "is approximately distributed as a unit normal." In addition, if we replace the unknown standard deviation σ by its estimator S, the sample standard deviation, then it still remains the case (by a result known as Slutsky's theorem) that the resulting quantity is approximately a unit normal. That is, when n is large

$$\sqrt{n}\,(\overline{X}-\theta)/S \mathrel{\dot\sim} N(0,\ 1) \tag{7.8}$$

Now for any α, $0 < \alpha < 1$, let z_α be such that

$$P\{Z > z_\alpha\} = \alpha$$

where Z is a unit normal random variable. (For example, $z_{.025} = 1.96$.) It follows from the symmetry of the unit normal density function about the origin that $z_{1-\alpha}$, the point at which the area under the density to its right is equal to $1 - \alpha$, is such that (see Figure 7.1)

$$z_{1-\alpha} = -z_\alpha$$

Therefore (see Figure 7.1)

$$P\{-z_{\alpha/2} < Z < z_{\alpha/2}\} = 1 - \alpha$$

It thus follows from (7.8) that

$$P\left\{-z_{\alpha/2} < \sqrt{n}\frac{(\overline{X}-\theta)}{S} < z_{\alpha/2}\right\} \approx 1 - \alpha$$

or, equivalently, upon multiplying by -1,

$$P\left\{-z_{\alpha/2} < \sqrt{n}\,\frac{(\theta-\overline{X})}{S} < z_{\alpha/2}\right\} \approx 1 - \alpha$$

which is equivalent to

$$P\left\{\overline{X} - z_{\alpha/2}\frac{S}{\sqrt{n}} < \theta < \overline{X} + z_{\alpha/2}\frac{S}{\sqrt{N}}\right\} \approx 1 - \alpha \tag{7.9}$$

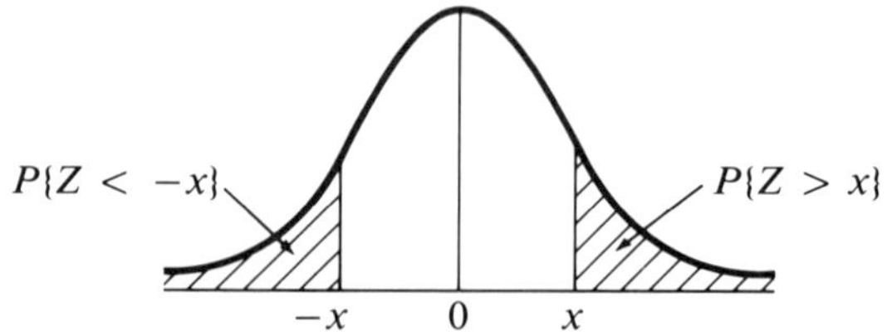

Figure 7.1 Unit normal density.

In other words, with probability $1 - \alpha$ the population mean θ will lie within the region $\overline{X} \pm z_{\alpha/2}S/\sqrt{n}$.

Definition:

If the observed values of the sample mean and the sample standard deviation are $\overline{X} = \overline{x}$ and $S = s$, *call the interval* $\overline{x} \pm z_{\alpha/2}s/\sqrt{n}$ *an (approximate)* $100(1 - \alpha)$ *percent confidence interval estimate of* θ.

REMARKS

1. To clarify the meaning of a "$100(1 - \alpha)$ percent confidence interval," consider, for example, the case where $\alpha = .05$, and so $z_{\alpha/2} = 1.96$. Now before the data are observed, it will be true, with probability (approximately) equal to .95, that the sample mean $\overline{X}$ and the sample standard deviation S will be such that θ will lie between $\overline{X} \pm 1.96S/\sqrt{n}$. After $\overline{X}$ and S are observed to equal, respectively, $\overline{x}$ and s, there is no longer any probability concerning whether θ lies in the interval $\overline{x} \pm 1.96s/\sqrt{n}$, for either it does or it does not. However, we are "95% confident" that in this situation it does lie in this interval (since we know that over the long run such intervals will indeed contain the mean 95 percent of the time).

2. (A technical remark). The above analysis is based on Equation (7.8), which states that $\sqrt{n}(\overline{X} - \theta)/S$ is approximately a unit normal random variable when n is large. Now if the original data values X_i were themselves normally distributed, then it is known that this quantity has (exactly) a t-distribution with $n - 1$ degrees of freedom. For this reason, many authors have proposed using this approximate distribution in the general case where the original distribution need not be normal. However, since it is not clear that the t-distribution with $n - 1$ degrees of freedom results in a better approximation than the normal in the general case, and because these two distributions are approximately equal for large n, we have used the normal approximation rather than introducing the t-random variable. ■

Consider now the case, as in a simulation study, where additional data values can be generated and the question is to determine when to stop generating new data values. One solution to this is to initially choose values α and l and to continue generating data until the approximate $100(1 - \alpha)$ percent confidence interval estimate of θ is less than l. Since the length of this interval will be $2z_{\alpha/2}S/\sqrt{n}$ we can accomplish this by the following technique.

1. Generate at least 30 data values.
2. Continue to generate additional data values, stopping when the number of values you have generated, call it k, is such that $2z_{\alpha/2}S/\sqrt{k} < l$, where S is the sample standard deviation based on those k values. (The value of S should be constantly updated, using the recursion given by (7.6) and (7.7), as new data are generated.)
3. If $\bar{x}$ and s are the observed values $\bar{X}$ and S, then the $100(1-\alpha)$ percent confidence interval estimate of θ, whose length is less than l, is $\bar{x} \pm 2z_{\alpha/2}s/\sqrt{k}$.

A Technical Remark

The more statistically sophisticated reader might wonder about our use of an approximate confidence interval whose theory was based on the assumption that the sample size was fixed when in the above situation the sample size is clearly a random variable depending on the data values generated. This, however, can be justified when the sample size is large and so from the viewpoint of simulation we can safely ignore this subtlety. ■

As noted in the previous section, the analysis is modified when $X_1, \ldots, X_n$ are Bernoulli random variables such that

$$X_i = \begin{cases} 1 & \text{with probability } p \\ 0 & \text{with probability } 1-p \end{cases}$$

Since in this case $\text{Var}(X_i)$ can be estimated by $\bar{X}(1-\bar{X})$, it follows that the equivalent statement to Equation (7.8) is that when n is large

$$\sqrt{n}\,\frac{(\bar{X}-p)}{\sqrt{\bar{X}(1-\bar{X})}} \mathrel{\dot\sim} N(0, 1) \tag{7.10}$$

Hence, for any α,

$$P\left\{-z_{\alpha/2} < \sqrt{n}\,\frac{(\bar{X}-p)}{\sqrt{\bar{X}(1-\bar{X})}} < z_{\alpha/2}\right\} = 1-\alpha$$

or, equivalently,

$$P\{\bar{X} - z_{\alpha/2}\sqrt{\bar{X}(1-\bar{X})/n} < p < \bar{X} + z_{\alpha/2}\sqrt{\bar{X}(1-\bar{X})/n}\} = 1-\alpha$$

Hence, if the observed value of $\bar{X}$ is p_n, we say that the "$100(1-\alpha)$ percent confidence interval estimate" of p is

$$p_n \pm z_{\alpha/2}\sqrt{p_n(1-p_n)/n}$$

7.3. The Bootstrapping Technique for Estimating Mean Square Errors

Suppose now that $X_1, \ldots, X_n$ are independent random variables having a common distribution function F, and suppose we are interested in using them to estimate some parameter $\theta(F)$ of the distribution F. For example, $\theta(F)$ could be (as in the previous sections of this chapter) the mean of F, or it could be the median or the variance of F, or any other parameter of F. Suppose further that an estimator of $\theta(F)$—call it $g(X_1, \ldots, X_n)$—has been proposed, and in order to judge its worth as an estimator of $\theta(F)$ we are interested in estimating its mean square error. That is, we are interested in estimating the value of

$$\text{MSE}(F) \equiv E_F[(g(X_1, \ldots, X_n) - \theta(F))^2]$$

[where our choice of notation MSE(F) suppresses the dependence on the estimator g, and where we have used the notation E_F to indicate that the expectation is to be taken under the assumption that the random variables all have distribution F]. Now whereas there is an immediate estimator of the above MSE—namely, S^2/n—when $\theta(F) = E[X_i]$ and $g(X_1, \ldots, X_n) = \overline{X}$, it is not at all apparent how it can be estimated otherwise. We now present a useful technique, known as the bootstrap technique, for estimating this mean square error.

To begin, note that if the distribution function F were known then we could theoretically compute the expected square of the difference between θ and its estimator; that is, we could compute the mean square error. However, after we observe the values of the n data points, we have a pretty good idea what the underlying distribution looks like. Indeed, suppose that the observed values of the data are $X_i = x_i$, $i = 1, \ldots, n$. We can now estimate the underlying distribution function F by the so-called empirical distribution function F_e, where $F_e(x)$, the estimate of $F(x)$, the probability that a datum value is less than or equal to x, is just the proportion of the n data values that are less than or equal to x. That is,

$$F_e(x) = \frac{\text{number of } i : X_i \leq x}{n}$$

Another way of thinking about F_e is that it is the distribution function of a random variable X_e which is equally likely to take on any of the n values x_i, $i = 1, \ldots, n$. (If the values x_i are not all distinct, then the above is to be interpreted to mean that X_e will equal the value x_i with a probability equal to the number of j such that $x_j = x_i$ divided by n; that is, if $n = 3$ and $x_1 = x_2 = 1$, $x_3 = 2$, then X_e is a random variable that takes on the value 1 with probability $\frac{2}{3}$ and 2 with probability $\frac{1}{3}$.)

Now if F_e is "close" to F, as it should be when n is large [indeed, the strong law of large numbers implies that with probability 1, $F_e(x)$ converges to $F(x)$ as $n \to \infty$, and another result, known as the Glivenko–Cantelli theorem, states that this convergence will, with probability 1, be uniform in x], then $\theta(F_e)$ will probably be close to $\theta(F)$—assuming that θ is, in some sense, a continuous function of the distribution—and MSE(F) should approximately be equal to

$$\text{MSE}(F_e) = E_{F_e}[(g(X_1, \ldots, X_n) - \theta(F_e))^2]$$

In the above expression the X_i are to be regarded as being independent random variables having distribution function F_e. The quantity MSE(F_e) is called the *bootstrap approximation to the mean square error MSE(F).*

To obtain a feel for the effectiveness of the bootstrap approximation to the mean square error, let us consider the one case where its use is not necessary—namely, when estimating the mean of a distribution by the sample mean $\overline{X}$. (Its use is not necessary in this case because there already is an effective way of estimating the mean square error $E[(\overline{X} - \theta)^2] = \sigma^2/n$—namely, by using the observed value of S^2/n.)

Example 7d Suppose we are interested in estimating $\theta(F) = E[X]$ by using the sample mean $\overline{X} = \sum_{i=1}^n X_i/n$. If the observed data are x_i, $i = 1, \ldots, n$, then the empirical distribution F_e puts weight $1/n$ on each of the points $x_1, \ldots, x_n$ (combining weights if the x_i are not all distinct). Hence the mean of F_e is $\theta(F_e) = \bar{x} = \sum_{i=1}^n x_i/n$; and thus the bootstrap estimate of the mean square error, call it MSE(F_e), is given by

$$\text{MSE}(F_e) = E_{F_e}\left[\left(\sum_{i=1}^n \frac{X_i}{n} - \bar{x}\right)^2\right]$$

where $X_1, \ldots, X_n$ are independent random variables each distributed according to F_e. Since

$$E_{F_e}\left[\sum_{i=1}^n \frac{X_i}{n}\right] = E_{F_e}[X] = \bar{x}$$

it follows that

$$\text{MSE}(F_e) = \text{Var}_{F_e}\left(\sum_{i=1}^n \frac{X_i}{n}\right)$$

$$= \frac{\text{Var}_{F_e}(X)}{n}$$

Now

$$\begin{aligned}\text{Var}_{F_e}(X) &= E_{F_e}[(X - E_{F_e}[X])^2] \\ &= E_{F_e}[(X - \bar{x})^2] \\ &= \frac{1}{n}\left[\sum_{i=1}^{n} (x_i - \bar{x})^2\right]\end{aligned}$$

and so

$$\text{MSE}(F_e) = \frac{\sum_{i=1}^{n} (x_i - \bar{x})^2}{n^2}$$

which compares quite nicely with S^2/n, the usual estimate of the mean square error. Indeed, because the observed value of S^2/n is $\Sigma_{i=1}^{n} (x_i - \bar{x})^2/[n(n - 1)]$, the bootstrap approximation is almost identical. ■

If the data values are $X_i = x_i$, $i = 1, \ldots, n$ then, as the empirical distribution function F_e puts weight $1/n$ on each of the points x_i, it is usually easy to compute the value of $\theta(F_e)$: for example, if the parameter of interest $\theta(F)$ was the variance of the distribution F, then $\theta(F_e) = \text{Var}_{F_e}(X) = \Sigma_{i=1}^{n} (x_i - \bar{x})^2/n$. To determine the bootstrap approximation to the mean square error we then have to compute

$$\text{MSE}(F_e) = E_{F_e}[(g(X_1, \ldots, X_n) - \theta(F_e))^2]$$

However, since the above expectation is to be computed under the assumption that $X_1, \ldots, X_n$ are independent random variables distributed according to F_e, it follows that the vector $(X_1, \ldots, X_n)$ is equally likely to take on any of the n^n possible values $(x_{i_1}, x_{i_2}, \ldots, x_{i_n})$, $i_j \in \{1, 2, \ldots, n\}$, $j = 1, \ldots, n$. Therefore

$$\text{MSE}(F_e) = \sum_{i_n} \cdots \sum_{i_1} \frac{[g(x_{i_1}, \ldots, x_{i_n}) - \theta(F_e)]^2}{n^n}$$

where each i_j goes from 1 to n; and so the computation of $\text{MSE}(F_e)$ requires, in general, summing n^n terms—an impossible task when n is large.

However, as we know, there is an effective way to approximate the average of a large number of terms, namely, by using simulation. Indeed, we could generate a set of n independent random variables $X_1^1, \ldots, X_n^1$ each having distribution function F_e and then set

$$Y_1 = [g(X_1^1, \ldots, X_n^1) - \theta(F_e)]^2$$

Next, we generate a second set $X_1^2, \ldots, X_n^2$ and compute

$$Y_2 = [g(X_1^2, \ldots, X_n^2) - \theta(F_e)]^2$$

and so on, until we have collected the variables $Y_1, Y_2, \ldots, Y_r$. Because these Y_i are independent random variables having mean $\text{MSE}(F_e)$, it follows that we can use their average $\sum_{i=1}^{r} Y_i/r$ as an estimate of $\text{MSE}(F_e)$.

REMARKS

1. It is quite easy to generate a random variable X having distribution F_e. Because such a random variable should be equally likely to be $x_1, \ldots, x_n$, just generate a random number U and set $X = x_I$, where $I = \text{Int}(nU) + 1$. (It is easy to check that this will still work even when the x_i are not all distinct.)

2. The above simulation allows us to approximate $\text{MSE}(F_e)$, which is itself an approximation to the desired $\text{MSE}(F)$. As such, it has been reported that roughly 100 simulation runs—that is, choosing $r = 100$—is usually sufficient. ■

The following example illustrates the use of the bootstrap in analyzing the output of a queueing simulation.

Example 7e Suppose in Example 7a we are interested in estimating the long-run average amount of time a customer spends in the system. That is, letting W_i be the amount of time the ith entering customer spends in the system, $i \geq 1$, we are interested in

$$\theta \equiv \lim_{n\to\infty} \frac{W_1 + W_2 + \cdots + W_n}{n}$$

To show that the above limit does indeed exist (note that the random variables W_i are neither independent nor identically distributed—why?), let N_i denote the number of customers that arrive on day i, and let

$$D_1 = W_1 + \cdots + W_{N_1}$$
$$D_2 = W_{N_1+1} + \cdots + W_{N_1+N_2}$$

and, in general, for $i > 2$,

$$D_i = W_{N_1+\cdots+N_{i-1}+1} + \cdots + W_{N_1+\cdots+N_i}$$

In words, D_i is the sum of the times in the system of all arrivals on day i. We can now express θ as

$$\theta = \lim_{m\to\infty} \frac{D_1 + D_2 + \cdots + D_m}{N_1 + N_2 + \cdots + N_m}$$

where the above follows because the ratio is just the average time in the system of all customers arriving in the first m days. Upon dividing numerator and denominator by m, we obtain

$$\theta = \lim_{m\to\infty} \frac{(D_1 + \cdots + D_m)/m}{(N_1 + \cdots + N_m)/m}$$

Now as each day follows the same probability law, it follows that the random variables $D_1, \ldots, D_m$ are all independent and identically distributed, as are the random variables $N_1, \ldots, N_m$. Hence, by the strong law of large numbers, it follows that the average of the first m of the D_i will, with probability 1, converge to their common expectation, with a similar statement being true for the N_i. Therefore we see that

$$\theta = \frac{E[D]}{E[N]}$$

where $E[N]$ is the expected number of customers to arrive in a day, and $E[D]$ is the expected sum of the times those customers spend in the system.

To estimate θ we can thus simulate the system over k days, collecting on the ith run the data N_i, D_i, where N_i is the number of customers arriving on day i and D_i is the sum of the times they spend in the system, $i = 1, \ldots, k$. Because the quantity $E[D]$ can then be estimated by

$$\overline{D} = \frac{D_1 + D_2 + \cdots + D_k}{k}$$

and $E[N]$ by

$$\overline{N} = \frac{N_1 + N_2 + \cdots + N_k}{k}$$

it follows that $\theta = E[D]/E[N]$ can be estimated by

$$\text{Estimate of } \theta = \frac{\overline{D}}{\overline{N}} = \frac{D_1 + \cdots + D_k}{N_1 + \cdots + N_k}$$

which, it should be noted, is just the average time in the system of all arrivals during the first k days.

To estimate

$$\text{MSE} = E\left[\left(\frac{\sum_{i=1}^{k} D_i}{\sum_{i=1}^{k} N_i} - \theta\right)^2\right]$$

we employ the bootstrap approach. Suppose the observed value of D_i, N_i is d_i, n_i, $i = 1, \ldots, k$. That is, suppose that the simulation resulted in n_i

arrivals on day i spending a total time d_i in the system. Thus the empirical joint distribution function of the random vector D, N puts equal weight on the k pairs d_i, n_i, $i = 1, \ldots, k$. That is, under the empirical distribution function we have

$$P_{F_e}\{D = d_i,\, N = n_i\} = \frac{1}{k}, \qquad i = 1, \ldots, k$$

Hence

$$E_{F_e}[D] = \bar{d} = \frac{\sum_{i=1}^{k} d_i}{k}, \qquad E_{F_e}[N] = \bar{n} = \frac{\sum_{i=1}^{k} n_i}{k}$$

and thus

$$\theta(F_e) = \frac{\bar{d}}{\bar{n}}$$

Hence

$$\mathrm{MSE}(F_e) = E_{F_e}\left[\left(\frac{\sum_{i=1}^{k} D_i}{\sum_{i=1}^{k} N_i} - \frac{\bar{d}}{\bar{n}}\right)^2\right]$$

where the above is to be computed under the assumption that the k pairs of random vectors D_i, N_i are independently distributed according to F_e.

Since an exact computation of $\mathrm{MSE}(F_e)$ would require computing the sum of k^k terms, we now perform a simulation experiment to approximate it. We generate k independent pairs of random vectors D_i^1, N_i^1, $i = 1, \ldots, k$, according to the empirical distribution function F_e, and then compute

$$Y_1 = \left(\frac{\sum_{i=1}^{k} D_i^1}{\sum_{i=1}^{k} N_i^1} - \frac{\bar{d}}{\bar{n}}\right)^2$$

We then generate a second set D_i^2, N_i^2 and compute the corresponding Y_2. This continues until we have generated the r values $Y_1, \ldots, Y_r$ (where $r = 100$ will suffice). The average of these r values, $\sum_{i=1}^{r} Y_i/r$, is then used to estimate $\mathrm{MSE}(F_e)$, which is itself our estimate of MSE, the mean square error of our estimate of the average amount of time a customer spends in the system. ■

REMARK: THE REGENERATIVE APPROACH

The above analysis assumed that each day independently followed the same probability laws. In certain applications, the same probability laws describe the system not over days of fixed lengths but rather over cycles whose length is random. For example, consider a queueing system in which customers arrive in accordance with a Poisson process, and suppose that the first customer arrives at time 0. If the random time T represents the next time that an arrival finds the system empty, then we say that the time from 0 to T constitutes the first cycle. The second cycle would be the time from T until the first time point after T that an arrival finds the system empty, and so on. It is easy to see, in most models, that the movements of the process over each cycle are independent and identically distributed. Hence, if we regard a cycle as being a "day," then all of the analysis given above remains valid. For example, θ, the average amount of time that a customer spends in the system, is given by $\theta = E[D]/E[N]$, where D is the sum of the times in the system of all arrivals in a cycle and N is the number of such arrivals. If we now generate k cycles, our estimate of θ is still $\sum_{i=1}^{k} D_i / \sum_{i=1}^{k} N_i$. In addition, the mean square error of this estimate can be approximated by using the bootstrap approach exactly as above.

The technique of analyzing a system by simulating "cycles," that is, random intervals during which the process follows the same probability law, is called the regenerative approach. ■

Problems

1. For any set of numbers $x_1, \ldots, x_n$, prove algebraically that

$$\sum_{i=1}^{n} (x_i - \bar{x})^2 = \sum_{i=1}^{n} x_i^2 - n\bar{x}^2$$

where $\bar{x} = \sum_{i=1}^{n} x_i/n$.

2. Give a probabilistic proof of the result of Problem 1, by letting X denote a random variable that is equally likely to take on any of the values $x_1, \ldots, x_n$; and then by applying the identity $\text{Var}(X) = E[X^2] - (E[X])^2$.

3. Write a program that uses the recursions given by Equations (7.6) and (7.7) to calculate the sample mean and sample variance of a data set.

4. Continue to generate unit normal random variables until you have generated n of them, where $n \geq 30$ is such that $S/\sqrt{n} < .1$; where S is the sample standard deviation of the n data values.

(a) How many normals do you think will be generated?

(b) How many normals did you generate?

(c) What is the sample mean of all the normals generated?
(d) What is the sample variance?
(e) Comment on the results of (c) and (d). Were they surprising?

5. Repeat Problem 4 with the exception that you now continue generating unit normals until $S/\sqrt{n} < .01$.

6. Estimate $\int_0^1 \exp(x^2)\, dx$ by generating random numbers. Generate at least 100 values and stop when the standard deviation of your estimator is less than .01.

7. Use simulation to estimate $E[N]$, where

$$N = \text{Minimum } \{n: \sum_{i=1}^{n} U_i > 1\}$$

where the U_i are independent uniform (0, 1) random variables. Stop when you are 95 percent certain that your estimator is correct to within $\pm.01$.

8. Use the approach that is presented in Example 3a of Chapter 3 to obtain an interval of size less than .1, which we can assert, with 95 percent confidence, contains π. How many runs were necessary?

9. Repeat Problem 8 when we want the interval to be no greater than .01.

10. To estimate θ, we generated 20 independent values having mean θ. If the successive values obtained were

102, 112, 131, 107, 114, 95, 133, 145, 139, 117
93, 111, 124, 122, 136, 141, 119, 122, 151, 143

how many additional random variables do you think we will have to generate if we want to be 99 percent certain that our final estimate of θ is correct to within ± 0.5?

11. Let $X_1, \ldots, X_n$ be independent identically distributed random variables having unknown mean μ. For given constants $a < b$, we are interested in estimating $p = P\{a < \sum_{i=1}^{n} X_i/n - \mu < b\}$.
(a) Explain how we can use the bootstrap approach to estimate p.
(b) Estimate p if $n = 10$ and the values of the X_i are 56, 101, 78, 67, 93, 87, 64, 72, 80, and 69. Take $a = -5$, $b = 5$.

In the following three problems $X_1, \ldots, X_n$ is a sample from a distribution whose variance is (the unknown) σ^2. We are planning to estimate σ^2 by the sample variance $S^2 = \sum_{i=1}^{n} (X_i - \overline{X})^2/(n-1)$, and we want to use the bootstrap technique to estimate $\text{Var}(S^2)$.

12. Explain why it suffices to consider $\text{Var}(S^2)$ rather than its mean square error. What if we wanted to estimate σ, the standard deviation?

13. If $n = 2$ and $X_1 = 1$ and $X_2 = 3$, what is the bootstrap estimate of $\text{Var}(S^2)$?

14. If $n = 15$ and the data are

$$5, 4, 9, 6, 21, 17, 11, 20, 7, 10, 21, 15, 13, 16, 8$$

approximate (by a simulation) the bootstrap estimate of $\text{Var}(S^2)$.

15. Consider a single server system in which potential customers arrive in accordance with a Poisson process having rate 4.0. A potential customer will only enter if there are three or fewer other customers in the system when he or she arrives. The service time of a customer is exponential with rate 4.2. No additional customers are allowed in after time $T = 8$. (All time units are per hour.) Develop a simulation study to estimate the average amount of time that an entering customer spends in the system. Using the bootstrap approach, estimate the mean square error of your estimator.

CHAPTER 8

Variance Reduction Techniques

Introduction

In a typical scenario for a simulation study, one is interested in determining θ, a parameter connected with some stochastic model. To estimate θ, the model is simulated to obtain, among other things, the output datum X which is such that $\theta = E[X]$. Repeated simulation runs, the ith one yielding the output variable X_i, are performed. The simulation study is then terminated when n runs have been performed and the estimate of θ is given by $\overline{X} = \Sigma_{i=1}^n X_i/n$. Because this results in an unbiased estimate of θ, it follows that its mean square error is equal to its variance. That is,

$$\text{MSE} = E[(\overline{X} - \theta)^2] = \text{Var}(\overline{X}) = \frac{\text{Var}(X)}{n}$$

Hence, if we can obtain a different unbiased estimate of θ having a smaller variance than does $\overline{X}$, we would obtain an improved estimator.

In this chapter we present a variety of different methods that one can attempt to use so as to reduce the variance of the (so-called raw) simulation estimate $\overline{X}$.

However, before presenting these variance reduction techniques, let us illustrate the potential pitfalls, even in quite simple models, of using the raw simulation estimator.

Example 8a: Quality Control. Consider a process that produces items sequentially. Suppose that these items have measurable values attached to them and that when the process is "in control" these values (suitably normalized) come from a unit normal distribution. Suppose further that when the process goes "out of control" the distribution of these values changes from the unit normal to some other distribution.

To help detect when the process goes out of control the following type of procedure, called an exponentially weighted moving-average control rule, is often used. Let $X_1, X_2, \ldots$ denote the sequence of data values. For a fixed value α, $0 \leq \alpha \leq 1$, define the sequence S_n, $n \geq 0$, by

$$S_0 = 0$$
$$S_n = \alpha S_{n-1} + (1 - \alpha)X_n, \qquad n \geq 1$$

Now when the process is in control, all the X_n have mean 0, and thus it is easy to verify that, under this condition, the exponentially weighted moving-average values S_n also have mean 0. The moving-average control rule is to fix a constant B, along with the value of α, and then to declare the process "out of control" when $|S_n|$ exceeds B. That is, the process is declared out of control at the random time N, where

$$N = \text{Min}\{n: \ |S_n| > B\}$$

Now it is clear that eventually $|S_n|$ will exceed B and so the process will be declared out of control even if it is still working properly—that is, even when the data values are being generated by a unit normal distribution. To make sure that this does not occur too frequently, it is prudent to choose α and B so that, when the X_n, $n \geq 1$, are indeed coming from a unit normal distribution, $E[N]$ is large. Suppose that it has been decided that, under these conditions, a value for $E[N]$ of 800 is acceptable. Suppose further that it is claimed that the values $\alpha = .9$ and $B = 0.8$ achieve a value of $E[N]$ of around 800. How can we check this claim?

One way of verifying the above claim is by simulation. Namely, we can generate unit normals X_n, $n \geq 1$, until $|S_n|$ exceeds 0.8 (where $\alpha = .9$ in the defining equation for S_n). If N_1 denotes the number of normals needed until this occurs, then, for our first simulation run, we have the output variable N_1. We then generate other runs, and our estimate of $E[N]$ is the average value of the output data obtained over all runs.

However, let us suppose that we want to be 99 percent confident that our estimate of $E[N]$, under the in-control assumption, is accurate to within $\pm$ 0.1. Hence, since 99 percent of the time a normal random variable is within $\pm$ 2.33 standard deviations of its mean (i.e., $z_{.005} = 2.33$), it follows that the number of runs needed—call it n—is such that

$$\frac{2.33\sigma_n}{\sqrt{n}} \approx 0.1$$

where σ_n is the sample standard deviation based on the first n data values. Now σ_n will approximately equal $\sigma(N)$, the standard deviation of N, and we now argue that this is approximately equal to $E[N]$. The argument runs as follows: since we are assuming that the process remains in control throughout, most of the time the value of the exponentially weighted moving average is near the origin. Occasionally, by chance, it gets large and approaches, in absolute value, B. At such times it may go beyond B and the run ends or there may be a string of normal data values which, after a short time, eliminate the fact that the moving average had been large (this is so because the old values of S_i are continually multiplied by .9 and so lose their effect). Hence, if we know that the process has not yet gone out of control by some fixed time k, then, no matter what the value of k, it would seem that the value of S_k is around the origin. In other words, it intuitively appears that the distribution of time until the moving average exceeds the control limits is approximately memoryless; that is, it is approximately an exponential random variable. But for an exponential random variable Y, $\text{Var}(Y) = (E[Y])^2$. Since the standard deviation is the square root of the variance, it thus seems intuitive that, when in control throughout, $\sigma(N) \approx E[N]$. Hence, if the original claim that $E[N] \approx 800$ is correct, the number of runs needed is such that

$$\sqrt{n} \approx 23.3 \times 800$$

or

$$n \approx (23.3 \times 800)^2 \approx 3.47 \times 10^8$$

In addition, because each run requires approximately 800 normal random variables (again assuming the claim is roughly correct), we see that to do this simulation would require approximately $800 \times 3.47 \times 10^8 \approx 2.77 \times 10^{11}$ normal random variables—a formidable task. ■

8.1. The Use of Antithetic Variables

Suppose we are interested in using simulation to estimate $\theta = E[X]$ and suppose we have generated X_1 and X_2, identically distributed random variables having mean θ. Then

$$\text{Var}\left(\frac{X_1 + X_2}{2}\right) = \tfrac{1}{4}[\text{Var}(X_1) + \text{Var}(X_2) + 2\,\text{Cov}(X_1, X_2)]$$

Hence it would be advantageous (in the sense that the variance would be reduced) if X_1 and X_2 rather than being independent were negatively correlated.

To see how we might arrange for X_1 and X_2 to be negatively correlated, suppose that X_1 is a function of m random numbers: that is, suppose that

$$X_1 = h(U_1, U_2, \ldots, U_m)$$

where $U_1, \ldots, U_m$ are m independent random numbers. Now if U is a random number—that is, U is uniformly distributed on (0, 1)—then so is $1 - U$. Hence the random variable

$$X_2 = h(1 - U_1, 1 - U_2, \ldots, 1 - U_m)$$

has the same distribution as X_1. In addition, since $1 - U$ is clearly negatively correlated with U, we might hope that X_2 might be negatively correlated with X_1; and indeed this result can be proved in the special case where h is a monotone (either increasing or decreasing) function of each of its coordinates. [This result follows from a more general result which states that two increasing (or decreasing) functions of a set of independent random variables are positively correlated. Both results are presented in the Appendix.] Hence, in this case, after we have generated $U_1, \ldots, U_m$ so as to compute X_1, rather than generating a new independent set of m random numbers, we do better by just using the set $1 - U_1, \ldots, 1 - U_m$ to compute X_2. In addition, it should be noted that we obtain a double benefit: namely, not only does our resulting estimator have smaller variance (at least when h is a monotone function), but we are also saved the time of generating a second set of random numbers.

Example 8b: Simulating the Reliability Function†. Consider a system of n components, each of which is either functioning or failed. Letting

$$s_i = \begin{cases} 1 & \text{if component } i \text{ works} \\ 0 & \text{otherwise} \end{cases}$$

we call $\mathbf{s} = (s_1, \ldots, s_n)$ the state vector. Suppose also that there is a nondecreasing function $\phi(s_1, \ldots, s_n)$ such that

$$\phi(s_1, \ldots, s_n) = \begin{cases} 1 & \text{if the system works under state vector } s_1, \ldots, s_n \\ 0 & \text{otherwise} \end{cases}$$

The function $\phi(s_1, \ldots, s_n)$ is called the structure function.

Some common structure functions are the following:

(a) *The series structure:* For the series structure

$$\phi(s_1, \ldots, s_n) = \operatorname*{Min}_i s_i$$

The series system works only if all its components function.

(b) *The parallel structure:* For the parallel structure

$$\phi(s_1, \ldots, s_n) = \operatorname*{Max}_i s_i$$

Hence the parallel system works if at least one of its components works.

(c) *The k-of-n system:* The structure function

† From S. Ross, *Introduction to Probability and Statistics for Engineers and Scientists*, New York: John Wiley & Sons, 1987, with permission.

$$\phi(s_1, \ldots, s_n) = \begin{cases} 1 & \text{if } \sum_{i=1}^{n} s_i \geq k \\ 0 & \text{otherwise} \end{cases}$$

is called a k-of-n structure function. Since $\Sigma_{i=1}^{n} s_i$ represents the number of functioning components, a k-of-n system works if at least k of the n components are working.

It should be noted that a series system is an n-of-n system, whereas a parallel system is a 1-of-n system.

(d) *The bridge structure:* A five-component system for which

$$\phi(s_1, s_2, s_3, s_4, s_5) = \text{Max}(s_1 s_3 s_5,\ s_2 s_3 s_4,\ s_1 s_4,\ s_2 s_5)$$

is said to have a bridge structure. Such a system can be represented schematically by Figure 8.1. The idea of the diagram is that the system functions if a signal can go, from left to right, through the system. The signal can go through any given node i provided that component i is functioning. We leave it as an exercise for the reader to verify the formula given for the bridge structure function.

Let us suppose now that the states of the components—call them S_i—$i = 1, \ldots, n$, are independent random variables such that

$$P\{S_i = 1\} = p_i = 1 - P\{S_i = 0\} \qquad i = 1, \ldots, n$$

Let

$$\begin{aligned} r(p_i, \ldots, p_n) &= P\{\phi(S_1, \ldots, S_n) = 1\} \\ &= E[\phi(S_1, \ldots, S_n)] \end{aligned}$$

The function $r(p_1, \ldots, p_n)$ is called the *reliability* function. It represents the probability that the system will work when the components are independent with component i functioning with probability p_i, $i = 1, \ldots, n$.

For a series system

$$r(p_1, \ldots, p_n) = P\{S_i = 1 \quad \text{for all } i = 1, \ldots, n\}$$

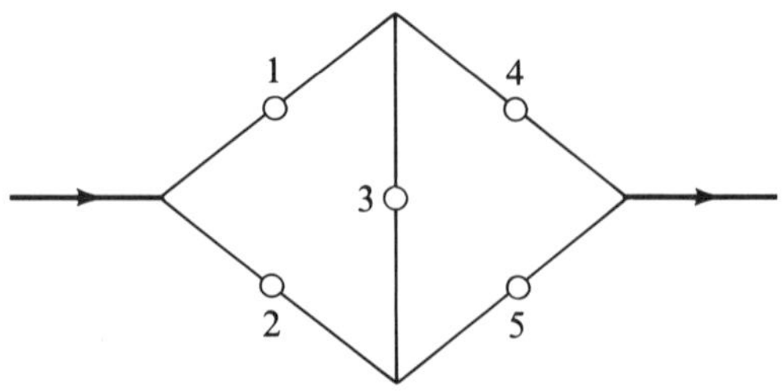

Figure 8.1 The bridge structure.

$$= \prod_{i=1}^{n} P\{S_i = 1\}$$

$$= \prod_{i=1}^{n} p_i$$

and for a parallel system

$$\begin{aligned} r(p_1, \ldots, p_n) &= P\{S_i = 1 \quad \text{for at least one } i,\ i = 1, \ldots, n\} \\ &= 1 - P\{S_i = 0 \quad \text{for all } i = 1, \ldots, n\} \\ &= 1 - \prod_{i=1}^{n} P(S_i = 0) \\ &= 1 - \prod_{i=1}^{n} (1 - p_i) \end{aligned}$$

However, for most systems it remains a formidable problem to compute the reliability function (even for such small systems as a 5-of-10 system or the bridge system it can be quite tedious to compute). So let us suppose that for a given nondecreasing structure function ϕ and given probabilities $p_1, \ldots, p_n$, we are interested in using simulation to estimate

$$r(p_1, \ldots, p_n) = E[\phi(S_1, \ldots, S_n)]$$

Now we can simulate the S_i by generating uniform random numbers $U_1, \ldots, U_n$ and then setting

$$S_i = \begin{cases} 1 & \text{if } U_i < p_i \\ 0 & \text{otherwise} \end{cases}$$

Hence we see that

$$\phi(S_1, \ldots, S_m) = h(U_1, \ldots, U_n)$$

where h is a decreasing function of $U_1, \ldots, U_n$. Therefore

$$\text{Cov}(h(\mathbf{U}), h(\mathbf{1} - \mathbf{U})) \leq 0$$

and so the antithetic variable approach of using $U_1, \ldots, U_n$ to generate both $h(U_1, \ldots, U_n)$ and $h(1 - U_1, \ldots, 1 - U_n)$ results in a smaller variance than if an independent set of random numbers was used to generate the second value of h. ■

Oftentimes the relevant output of a simulation is a function of the input random variables $Y_1, \ldots, Y_m$. That is, the relevant output is $X =$

$h(Y_1, \ldots, Y_m)$. Suppose the Y_i are independent with Y_i having distribution F_i, $i = 1, \ldots, m$. If these input variables are generated by the inverse transform technique, we can write

$$X = h(F_1^{-1}(U_1), \ldots, F_m^{-1}(U_m))$$

where $U_1, \ldots, U_m$ are independent random numbers. Since a distribution function is increasing, it follows that its inverse is also increasing and thus if $h(y_1, \ldots, y_m)$ were a monotone function of its coordinates, then it follows that $h(F_1^{-1}(U_1), \ldots, F_m^{-1}(U_m))$ will be a monotone function of the U_i. Hence the method of antithetic variables, which would first generate $U_1, \ldots, U_m$ to compute X_1 and then use $1 - U_1, \ldots, 1 - U_m$ to compute X_2, would result in an estimator having a smaller variance than would have been obtained if a new set of random numbers were used for X_2.

Example 8c: Simulating a Queueing System. Consider a given queueing system, and let D_i denote the delay in queue of the ith arriving customer, and suppose we are interested in simulating the system so as to estimate $\theta = E[X]$, where

$$X = D_1 + \cdots + D_n$$

is the sum of the delays in queue of the first n arrivals. Let $I_1, \ldots, I_n$ denote the first n interarrival times (i.e., I_j is the time between the arrivals of customers $j - 1$ and j), and let $S_1, \ldots, S_n$ denote the first n service times of this system, and suppose that these random variables are all independent. Now in many systems X is a function of the $2n$ random variables $I_1, \ldots, I_n, S_1, \ldots, S_n$, say,

$$X = h(I_1, \ldots, I_n, S_1, \ldots, S_n)$$

Also, as the delay in queue of a given customer usually increases (depending of course on the specifics of the model) as the service times of other customers increase and usually decreases as the times between arrivals increase, it follows that, for many models, h is a monotone function of its coordinates. Hence, if the inverse transform method is used to generate the random variables $I_1, \ldots, I_n, S_1, \ldots, S_n$, then the antithetic variable approach results in a smaller variance. That is, if we initially use the $2n$ random numbers U_i, $i = 1, \ldots, 2n$, to generate the interarrival and service times by setting $I_i = F_i^{-1}(U_i)$, $S_i = G_i^{-1}(U_{n+i})$, where F_i and G_i are, respectively, the distribution functions of I_i and S_i, then the second simulation run should be done, in the same fashion, but using the random numbers $1 - U_i$, $i = 1, \ldots, 2n$. This results in a smaller variance than if a new set of $2n$ random numbers were generated for the second run. ■

The following example illustrates the sort of improvement that can sometimes be gained by the use of antithetic variables.

Example 8d: Suppose we were interested in using simulation to estimate

$$\theta = E[e^U] = \int_0^1 e^x \, dx$$

(Of course, we know that $\theta = e - 1$; however, the point of this example is to see what kind of improvement is possible by using antithetic variables.) Since the function $h(u) = e^u$ is clearly a monotone function, the antithetic variable approach leads to a variance reduction, whose value we now determine. To begin, note that

$$\begin{aligned}\mathrm{Cov}(e^U, e^{1-U}) &= E[e^U e^{1-U}] - E[e^U]E[e^{1-U}] \\ &= e - (e - 1)^2 = -.2342\end{aligned}$$

Also, because

$$\begin{aligned}\mathrm{Var}(e^U) &= E[e^{2U}] - (E[e^U])^2 \\ &= \int_0^1 e^{2x} \, dx - (e - 1)^2 \\ &= \frac{[e^2 - 1]}{2} - (e - 1)^2 = .2420\end{aligned}$$

we see that the use of independent random numbers results in a variance of

$$\mathrm{Var}\left(\frac{\exp\{U_1\} + \exp\{U_2\}}{2}\right) = \frac{\mathrm{Var}(e^U)}{2} = .1210$$

whereas the use of the antithetic variables U and $1 - U$ gives a variance of

$$\mathrm{Var}\left(\frac{e^U + e^{1-U}}{2}\right) = \frac{\mathrm{Var}(e^U)}{2} + \frac{\mathrm{Cov}(e^U, e^{1-U})}{2} = .0039$$

a variance reduction of 96.7 percent. ■

In the case of a normal random variable having mean μ and variance σ^2, we can use the antithetic variable approach by first generating such a random variable Y and then taking as the antithetic variable $2\mu - Y$, which is also normal with mean μ and variance σ^2 and is clearly negatively correlated with Y. If we were using simulation to compute $E[h(Y_1, \ldots, Y_n)]$, where the Y_i are independent normal random variables and h is a monotone function of its coordinates, then the antithetic approach of first generating the n normals $Y_1, \ldots, Y_n$ to compute $h(Y_1, \ldots, Y_n)$ and then using the antithetic variables $2\mu_i - Y_i$, $i = 1, \ldots, n$, to compute the next simulated value of h would lead to a reduction in variance as compared with generating a second set of n normal random variables.

8.2. The Use of Control Variates

Again suppose that we want to use simulation to estimate $\theta = E[X]$, where X is the output of a simulation. Now suppose that for some other output variable Y, the expected value of Y is known—say, $E[Y] = \mu_y$. Then for any constant c, the quantity

$$X + c(Y - \mu_y)$$

is also an unbiased estimator of θ. To determine the best value of c, note that

$$\begin{aligned}\text{Var}(X + c(Y - \mu_y)) &= \text{Var}(X + cY)\\ &= \text{Var}(X) + c^2\,\text{Var}(Y) + 2c\,\text{Cov}(X, Y)\end{aligned}$$

Simple calculus now shows that the above is minimized when $c = c^*$, where

$$c^* = -\frac{\text{Cov}(X, Y)}{\text{Var}(Y)} \tag{8.1}$$

and for this value the variance of the estimator is

$$\text{Var}(X + c^*(Y - \mu)) = \text{Var}(X) - \frac{[\text{Cov}(X, Y)]^2}{\text{Var}(Y)} \tag{8.2}$$

The quantity Y is called a *control variate* for the simulation estimator X. To see why it works, note that c^* is negative (positive) when X and Y are positively (negatively) correlated. So suppose that X and Y were positively correlated, meaning, roughly, that X is large when Y is large and vice versa. Hence, if a simulation run results in a large (small) value of Y—which is indicated by Y being larger than its known mean μ_y—then it is probably true that X is also larger (smaller) than its mean θ, and so we would like to correct for this by lowering (raising) the value of the estimator X, and this is done since c^* is negative (positive). A similar argument holds when X and Y are negatively correlated.

While Var(Y) may be known in advance of the simulation, it is almost always the case that Cov(X, Y) is unknown. (Indeed, the whole point of the simulation is to estimate the unknown value $E[X]$.) Hence the value c^* is not known before the simulation experiment. However, because most of the maximal gain involved in using the control variate is obtained by using a value close to c^*, we can still effect a substantial improvement over the raw estimator X by using the result of the simulation to estimate c^*. Indeed, suppose we perform n simulation runs with output variables X_i, Y_i from the ith run, $i = 1, \ldots, n$. Then

$$\text{Cov}(X, Y) = E[(X - E[X])(Y - E[Y])]$$

can best be estimated by

$$\text{Estimate of } \mathrm{Cov}(X, Y) = \frac{1}{n} \sum_{i=1}^{n} (X_i - \overline{X})(Y_i - \mu_y) \tag{8.3}$$

where $\overline{X}$ and $\overline{Y}$ are, respectively, the sample means of the X's and the Y's. The value of c^* should now be estimated by the negative of (8.3) divided either by $\mathrm{Var}(Y)$, if that quantity is known, or by its estimator, $\sum_{i=1}^{n}(Y_i - \overline{Y})^2/(n-1)$, otherwise. Calling this estimator c_e^*, the estimator of $\theta = E[X]$ should now be $\overline{X} + c_e^* (\overline{Y} - \mu y)$.

Example 8e: Suppose, as in example 8b, that we wanted to use simulation to estimate the reliability function

$$r(p_1, \ldots, p_n) = E[\phi(S_1, \ldots, S_n)]$$

where

$$S_i = \begin{cases} 1 & \text{if } U_i < p_i \\ 0 & \text{otherwise} \end{cases}$$

Since $E[S_i] = p_i$, it follows that

$$E\left[\sum_{i=1}^{n} S_i\right] = \sum_{i=1}^{n} p_i$$

Hence we can use the number of working components, $Y \equiv \sum S_i$, as a control variate of the estimator $X \equiv \phi(S_1, \ldots, S_n)$. Since $\sum_{i=1}^{n} S_i$ and $\phi(S_1, \ldots, S_n)$ are both increasing functions of the S_i, it is intuitive (and can be proved) that they are positively correlated; and thus the sign of c^* is negative. To estimate c^*, we need to first simulate and then use the resulting data to estimate $\mathrm{Cov}(X, Y)$. Because $\mathrm{Var}(Y)$ is known and is given by

$$\begin{aligned} \mathrm{Var}(Y) &= \mathrm{Var}\left(\sum_{i=1}^{n} S_i\right) \\ &= \sum_{i=1}^{n} \mathrm{Var}(S_i) \qquad \text{(by independence)} \\ &= \sum_{i=1}^{n} p_i(1 - p_i) \qquad (S_i \text{ is a Bernoulli random variable}) \end{aligned}$$

it need not be estimated by the simulation. ■

Example 8f: Consider a queueing system in which customers arrive in accordance with a nonhomogeneous Poisson process with intensity function $\lambda(s)$, $s > 0$. Suppose that the service times are independent random variables having distribution G and are also independent of the arrival times. Suppose we were interested in estimating the total time spent in the system by all customers arriving before time t. That is, if we let W_i denote the amount of time that the ith entering customer spends in the system, then we are interested in $\theta = E[X]$, where

$$X = \sum_{i=1}^{N(t)} W_i$$

and where $N(t)$ is the number of arrivals by time t. A natural quantity to use as a control in this situation is the total of the service times of all these customers. That is, let S_i denote the service time of the ith customer and set

$$Y = \sum_{i=1}^{N(t)} S_i$$

Since the service times are independent of $N(t)$, it follows that

$$E[Y] = E[S]E[N(t)]$$

where $E[S]$, the mean service time, and $E[N(t)]$, the mean number of arrivals by t, are both known quantities. ■

Example 8g: As in Example 8d, suppose we were interested in using simulation to compute $\theta = E[e^U]$. Here, a natural variate to use as a control is the random number U. To see what sort of improvement over the raw estimator is possible, note that

$$\begin{aligned}\text{Cov}(e^U, U) &= E[Ue^U] - E[U]E[e^U] \\ &= \int_0^1 xe^x\,dx - \frac{(e-1)}{2} \\ &= 1 - \frac{(e-1)}{2} = .14086\end{aligned}$$

Because $\text{Var}(U) = \frac{1}{12}$ it follows from (8.2) that

$$\begin{aligned}\text{Var}(e^U + c^*(U - \tfrac{1}{2})) &= \text{Var}(e^U) - 12(.14086)^2 \\ &= .2420 - .2380 = .0039\end{aligned}$$

where the above used, from Example 8d, that $\text{Var}(e^U) = .2420$. Hence, in this case, the use of the control variate U can lead to a variance reduction of up to 98.4 percent.

The optimal control constant c in the above is c^*, where

$$c^* = -\frac{\text{Cov}(e^U, U)}{\text{Var}(U)} = -12(.14086) = -1.6903$$

However, any control constant near c^* also yields a substantial variance reduction. For example, if we use the estimator

$$\text{Est} = e^U - 1.6(U - \tfrac{1}{2})$$

then

$$\begin{aligned}\text{Var(Est)} &= \text{Var}(e^U) + (1.6)^2\,\text{Var}(U) - 3.2\,\text{Cov}(e^U, U)\\ &= .2420 + .21333 - .45075 = .0046\end{aligned}$$

which, although not as low as the minimal possible variance of .0039, is still a substantial improvement over the uncontrolled estimator whose variance is given by $\text{Var}(e^U) = .2420$. (Indeed, it is a variance reduction of 98.1 percent in comparison with the raw estimator.) ■

Example 8h: A List Reordering Problem. Suppose we are given a set of n elements, numbered 1 through n, which are to be arranged in an ordered list. At each unit of time a request is made to retrieve one of these elements, with the request being for element i with probability $p(i)$, $\Sigma_{i=1}^{n} p(i) = 1$. After being requested, the element is put back in the list but not necessarily in the same position. For example, a common reordering rule is to interchange the requested element with the one immediately preceding it. Thus if $n = 4$ and the present ordering is 1, 4, 2, 3, then under this rule a request for element 2 would result in the reorder 1, 2, 4, 3. Starting with an initial ordering that is equally likely to be any of the $n!$ orderings and using this interchange rule, suppose we are interested in determining the expected sum of the positions of the first N elements requested. How can we efficiently accomplish this by simulation?

One effective way is as follows. The "natural" way of simulating the above is first to generate a random permutation of 1, 2, . . . , n to establish the initial ordering, and then at each of the next N periods determine the element requested by generating a random number U and then letting the request be for element j if $\Sigma_{k=1}^{j-1} p(k) < U \le \Sigma_{k=1}^{j} p(k)$. However, a better technique is to generate the element requested in such a way so that small values of U correspond to elements close to the front. Specifically, if the present ordering is $i_1, i_2, \ldots, i_n$, then generate the element requested by generating a random number U and then letting the selection be for i_j if $\Sigma_{k=1}^{j-1} p(i_k) < U \le \Sigma_{k=1}^{j} p(i_k)$. For example, if $n = 4$ and the present ordering is 3, 1, 2, 4, then we should generate U and let the selection be for 3 if $U \le p(3)$, let it be for 1 if $p(3) < U \le p(3) + p(1)$, and so on. As small values of U thus correspond to elements near the front, we can use $\Sigma_{r=1}^{N} U_r$ as a control variable, where U_r is the random number used for the rth request in a run. That is, if P_r is the position of the rth selected element in a run, then rather than just using the raw estimator $\Sigma_{r=1}^N P_r$ we should use

$$\sum_{r=1}^{N} P_r + c^*\left(\sum_{r=1}^{N} U_r - \frac{N}{2}\right)$$

where

$$c^* = -\frac{\text{Cov}\left(\sum_{r=1}^{N} P_r, \sum_{r=1}^{N} U_r\right)}{\frac{N}{12}}$$

and where the above covariance should be estimated using the data from all the simulated runs. ■

Of course, one can use more than a single variable as a control. For example, if a simulation results in output variables Y_i, $i = 1, \ldots, k$, and $E[Y_i] = \mu_i$ is known, then for any constants c_i, $i = 1, \ldots, k$, we may use

$$X + \sum_{i=1}^{k} c_i(Y_i - \mu_i)$$

as an unbiased estimator of $E[X]$.

Remarks

1. The maximal benefit when using an estimator of the form $X + c(Y - \mu_y)$ is obtained when $c = c^*$ as defined in Equation (8.1). However, there is very little lost if one uses any value c close to c^*. Hence, from a practical point of view, to reduce the amount of computation necessary, it is probably best to use some initial runs (say, the first few hundred) to estimate c^*, say, by c_e^*. We can then fix c at this value and use the output variables $X_i + c_e^*(Y_i - \mu_y)$ from all the runs generated. The statistical analysis involved in determining when to stop and the determination of confidence intervals are now exactly as detailed in Chapter 7 with the use of these new output variables. (Fixing the value of c^* by using only some of our runs saves us the work of continually updating our estimates of c^*.)

2. A valuable way of interpreting the control variable approach is that it combines estimators of θ. That is, suppose the values of X and W are both determined by the simulation, and suppose $E[X] = E[W] = \theta$. Then we may consider any unbiased estimator of the form

$$\alpha X + (1 - \alpha)W$$

The best such estimator, which is obtained by choosing α to minimize the variance, is given by letting $\alpha = \alpha^*$, where

$$\alpha^* = \frac{\mathrm{Var}(W) - \mathrm{Cov}(X, W)}{\mathrm{Var}(X) + \mathrm{Var}(W) - 2\,\mathrm{Cov}(X, W)} \tag{8.4}$$

Now if $E[Y] = \mu_y$ is known, we have the two unbiased estimators X and $X + Y - \mu_y$. The combined estimator can then be written as

$$(1 - c)X + c(X + Y - \mu_y) = X + c(Y - \mu_y)$$

To go the other way in the equivalence between control variates and combining estimators, suppose that $E[X] = E[W] = \theta$. Then if we use X, controlling with the variable $Y = X - W$, which is known to have mean 0, we then obtain an estimator of the form

$$X + c(X - W) = (1 + c)X - cW$$

which is a combined estimator with $\alpha = 1 + c$.

3. With the interpretation given in Remark 2, the antithetic variable approach may be regarded as a special case of control variables. That is, if $E[X] = \theta$, where $X = h(U_1, \ldots, U_n)$, then also $E[W] = \theta$, where $W = h(1 - U_1, \ldots, 1 - U_n)$. Hence we can combine to get an estimator of the form $\alpha X + (1 - \alpha)W$. Since $\mathrm{Var}(X) = \mathrm{Var}(W)$, as X and W have the same distribution, it follows from Equation (8.4) that the best value of α is $\alpha = \frac{1}{2}$, and this is the antithetic variable estimator. ■

8.3. Variance Reduction by Conditioning

Recall the conditional variance formula proved in Section 2.10 of Chapter 2.

$$\mathrm{Var}(X) = E[\mathrm{Var}(X|Y)] + \mathrm{Var}(E[X|Y])$$

Since both terms on the right are nonnegative, because a variance is always nonnegative, we see that

$$\mathrm{Var}(X) \geq \mathrm{Var}(E[X|Y]) \tag{8.5}$$

Now suppose we are interested in performing a simulation study so as to ascertain the value of $\theta = E[X]$, where X is an output variable of a simulation run. Also, suppose there is a second variable Y, such that $E[X|Y]$ is known and takes on a value that can be determined from the simulation run. Since

$$E[E[X|Y]] = E[X] = \theta$$

it follows that $E[X|Y]$ is also an unbiased estimator of θ; and thus using (8.5) it follows that $E[X|Y]$ is a superior estimator of θ than is the (raw) estimator X.

REMARK

To understand why the conditional expectation estimator is superior to the raw estimator, note first that we are performing the simulation to estimate the unknown value of $E[X]$. We can now imagine that a simulation run proceeds in two stages: first, we observe the simulated value of the random variable Y and then the simulated value of X. However, after observing Y if we are now able to compute the (conditional) expected value of X, then by using this value we obtain an estimate of $E[X]$, which eliminates the additional variance involved in simulating the actual value of X.

We now illustrate the uses of "conditioning" by a series of examples.

Example 8i: Let us reconsider our use of simulation to estimate π. In Example 3a of Chapter 3, we showed how we can estimate π by determining how often a randomly chosen point in the square of area 4 centered around the origin falls within the inscribed circle of radius 1. Specifically, if we let $V_i = 2U_i - 1$, where U_i, $i = 1, 2$, are random numbers, and set

$$I = \begin{cases} 1 & \text{if } V_1^2 + V_2^2 \leq 1 \\ 0 & \text{otherwise} \end{cases}$$

then, as noted in Example 3a, $E[I] = \pi/4$.

The use of the average of successive values of I to estimate $\pi/4$ can be improved upon by using $E[I|V_1]$ rather than I. Now

$$\begin{aligned} E[I|V_1 = v] &= P\{V_1^2 + V_2^2 \leq 1 | V_1 = v\} \\ &= P\{v^2 + V_2^2 \leq 1 | V_1 = v\} \\ &= P\{V_2^2 \leq 1 - v^2\} \qquad \text{by the independence of } V_1 \text{ and } V_2 \\ &= P\{-(1 - v^2)^{1/2} \leq V_2 \leq (1 - v^2)^{1/2}\} \\ &= \int_{-(1-v^2)^{1/2}}^{(1-v^2)^{1/2}} \left(\frac{1}{2}\right) dx \qquad \text{since } V_2 \text{ is uniform over } (-1, 1) \\ &= (1 - v^2)^{1/2} \end{aligned}$$

Hence

$$E[I|V_1] = (1 - V_1^2)^{1/2}$$

and so the estimator $(1 - V_1^2)^{1/2}$ also has mean $\pi/4$ and has a smaller variance than I. Since

$$\begin{aligned} E[(1 - V_1^2)^{1/2}] &= \int_{-1}^{1} (1-x^2)^{1/2} \left(\frac{1}{2}\right) dx \\ &= \int_0^1 (1 - x^2)^{1/2}\, dx \\ &= E[(1 - U^2)^{1/2}] \end{aligned}$$

we can simplify somewhat by using the estimator $(1 - U^2)^{1/2}$, where U is a random number.

The improvement in variance obtained by using the estimator $(1 - U^2)^{1/2}$ over the estimator I is easily determined.

$$\mathrm{Var}[(1 - U^2)^{1/2}] = E[1 - U^2] - \left(\frac{\pi}{4}\right)^2$$

$$= \frac{2}{3} - \left(\frac{\pi}{4}\right)^2 = .0498$$

where the first inequality used the identity $\mathrm{Var}(W) = E[W^2] - (E[W])^2$. On the other hand, because I is a Bernoulli random variable having mean $\pi/4$, we have

$$\mathrm{Var}(I) = \left(\frac{\pi}{4}\right)\left(1 - \frac{\pi}{4}\right) = .1686$$

thus showing that conditioning results in a 70.44 percent reduction in variance. (In addition, only one rather than two random numbers is needed for each simulation run, although a computational cost of having to compute a square root must be paid.)

Since the function $(1 - u^2)^{1/2}$ is clearly a monotone decreasing function of u in the region $0 < u < 1$, it follows that the estimator $(1 - U^2)^{1/2}$ can be improved upon by using antithetic variables. That is, the estimator

$$\tfrac{1}{2}[(1 - U^2)^{1/2} + (1 - (1 - U)^2)^{1/2}]$$

has smaller variance than $\frac{1}{2}[(1 - U_1^2)^{1/2} + (1 - U_2^2)^{1/2}]$.

Another way of improving the estimator $(1 - U^2)^{1/2}$ is by using a control variable. A natural control variable in this case is U^2 and, because $E[U^2] = \frac{1}{3}$, we could use an estimator of the type

$$(1 - U^2)^{1/2} + c(U^2 - \tfrac{1}{3})$$

The best c—namely, $c^* = -\mathrm{Cov}[(1 - U^2)^{1/2}, U^2)/\mathrm{Var}(U^2)$, can be estimated by using the simulation to estimate the covariance term. (We could also have tried to use U as a control variable; it makes a difference since a correlation between two random variables is only a measure of their "linear dependence" rather than of their total dependence. But the use of U^2 leads to a greater improvement; see Problem 14.) ■

Example 8j: The Reliability Function Revisited. Suppose, as in Example 8b, that S_j, $j = 1, \ldots, n$, are independent with

$$P\{S_j = 1\} = 1 - P\{S_j = 0\} = p_j$$

and suppose we are interested in estimating the reliability function $E[\phi(S_1, \ldots, S_n)]$, where ϕ is an increasing binary (i.e., its only possible values are 0 and 1) function. If we generate the values of $S_1, \ldots, S_n$, then an improvement over the raw estimator $\phi(S_1, \ldots, S_n)$ is to take its conditional expectation given all of the S_j except one. That is, for fixed i,

$$E[\phi(S_1, \ldots, S_n) \mid S_1, \ldots, S_{i-1}, S_{i+1}, \ldots, S_n] \tag{8.6}$$

will be an improved estimator. (To use the above estimator we would, on any given simulation run, simulate all the component states S_j except for S_i. The estimate for that simulation run will take on one of three possible values: either it will equal 1 if, given the states S_j, $j \neq i$, the system would function even if component i did not function; or it will equal 0 if the system will not function even if component i did function; or it will equal p_i if the system will function if i functions and would be failed otherwise.)

The above estimator, given by (8.6), can now be further improved upon either by using antithetic variables—since the estimator is still a monotone function—or by controlling on $\Sigma_{j \neq 1} S_j$ the number of the simulated components that work. ■

Example 8k: A Finite Capacity Queueing Model. Consider a queueing system in which arrivals enter only if there are fewer than N other customers in the system when they arrive. Any customer encountering N others upon arrival is deemed to be lost to the system. Suppose further that potential customers arrive in accordance with a Poisson process having rate λ; and suppose we are interested in using simulation to estimate the expected number of lost customers by some fixed time t.

A simulation run would consist of simulating the above system up to time t. If, for a given run, we let L denote the number of lost customers, then the average value of L, over all simulation runs, is the (raw) simulation estimator of the desired quantity $E[L]$. However, we can improve upon this estimator by conditioning upon the amount of time that the system is at capacity. That is, rather than using L, the actual number of lost customers up to time t, we consider $E[L|T_C]$, where T_C is the total amount of time in the interval $(0, t)$ that there are N customers in the system. Since customers are always arriving at the Poisson rate λ no matter what is happening within the system, it follows that

$$E[L|T_C] = \lambda T_C$$

Hence an improved estimator is obtained by ascertaining, for each run, the total time in which there are N customers in the system—say, $T_{C,i}$ is the time at capacity during the ith run. Then the improved estimator of $E[L]$ is $\lambda \ \Sigma_{i=1}^{k} T_{C,i}/k$, where k is the number of simulation runs. (In effect, since the expected number of lost customers given the time at capacity T_C is just λT_C, what this estimator does is use the actual conditional expectation rather than simulating (and increasing the variance of the estimator) a Poisson random variable having this mean.

If the arrival process were a nonhomogeneous Poisson process having intensity function $\lambda(s)$, $0 \leq s \leq t$, then we would not be able to compute the conditional expected number of lost customers if we were given only the total time at capacity. What we now need is the actual times at which the system was at capacity. So let us condition on the intervals during which the system was at capacity. Now letting N_C denote the number of intervals during $(0, t)$ during which the system is at capacity, and letting those intervals be designated by $I_1, \ldots, I_{N_C}$, then

$$E[L|N_C, I_1, \ldots, I_{N_C}] = \sum_{i=1}^{N_C} \int_{I_i} \lambda(s)\, ds$$

The use of the average value, over all simulation runs, of the above quantity leads to a better estimator—in the sense of having a smaller mean square error—of $E[L]$ than would the raw simulation estimator of the average number lost per run.

The above estimators can be further improved by using a control variate. One that quickly comes to mind would be to control using the service times. That is, if the model supposed that all service times were independent of the arrival process and had mean μ, then we could control with the zero-mean random quantity

$$\sum_{i=1}^{M} (S_i - \mu)$$

where M is the total number of customers to have entered (although not necessarily completed) service by time t and S_i is the ith service time. ∎

Example 8l: Suppose we wanted to estimate the expected sum of the times in the system of the first n customers in a queueing system. That is, if W_i is the time that the ith customer spends in the system, we are interested in estimating

$$\theta = E\left[\sum_{i=1}^{n} W_i\right]$$

Let S_i denote the "state of the system" at the moment that the ith customer arrives, and consider the estimator

$$\sum_{i=1}^{n} E[W_i \mid S_i]$$

Since

$$E\left[\sum_{i=1}^{n} E[W_i \mid S_i]\right] = \sum_{i=1}^{n} E[E[W_i \mid S_i]] = \sum_{i=1}^{n} E[W_i] = \theta$$

it follows that this is an unbiased estimator of θ. It has recently been shown[†] that, in a wide class of models, this estimator has a smaller variance than the raw simulation estimator $\Sigma_{i=1}^{n} W_i$. (It should be noted that whereas it is immediate that $E[W_i \mid S_i]$ has smaller variance than W_i, this does not imply, because of the covariance terms, that $\Sigma_{i=1}^{n} E[W_i \mid S_i]$ has smaller variance than $\Sigma_{i=1}^{n} W_i$.)

The quantity S_i, which refers to the state of the system as seen by the ith customer upon its arrival, is supposed to represent the least amount of information that enables us to compute the conditional expected time that the customer spends in the system. For example, if there is a single server and the service times are all exponential with mean μ, then S_i would refer to N_i, the number of customers in the system encountered by the ith arrival. In this case,

$$E[W_i \mid S_i] = E[W_i \mid N_i] = (N_i + 1)\mu$$

which follows since the ith arrival will have to wait for N_i service times (one of which is the completion of service of the customer presently being served when customer i arrives—but, by the memoryless property of the exponential, that remaining time will also be exponential with mean μ) all having mean μ and then to this we must add its own service time. Thus the estimator that takes the average value, over all simulation runs, of the quantity $\Sigma_{i=1}^{n}(N_i + 1)\mu$ is a better estimator than the average value of $\Sigma_{i=1}^{n} W_i$. ■

Estimating the Expected Number of Renewals by Time t

Suppose that "events" are occurring randomly in time. Let T_1 denote the time of the first event, T_2 the time between the first and second event, and, in general, T_n the time between the $(n - 1)$st and the nth event, $n \geq 1$. If we let

$$S_n = \sum_{i=1}^{n} T_i$$

the first event occurs at time S_1, the second at time S_2, and, in general, the nth event occurs at time S_n (see Figure 8.2). Let $N(t)$ denote the number of events that occur by time t; that is, $N(t)$ is the largest n for which the nth event occurs by time t, or, equivalently,

[†] S. M. Ross, "Simulating Average Delay—Variance Reduction by Conditioning," *Probability in the Engineering and Informational Sciences,* Vol. 2, No. 3, 1988.

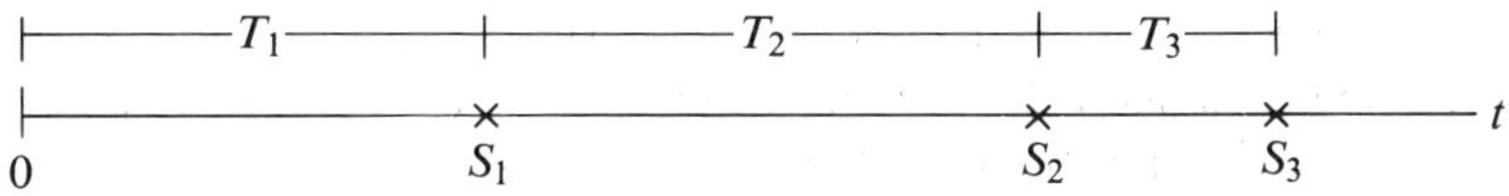

Figure 8.2 x = event.

$$N(t) = \text{Max}\{n: S_n \leq t\}$$

If the interevent times $T_1, T_2, \ldots$ are independent and identically distributed according to some distribution function F, then the process $\{N(t), t \geq 0\}$ is called a *renewal process*.

A renewal process is easily simulated by generating the interarrival times. Suppose now that we wanted to use simulation to estimate $\theta = E[N(t)]$, the mean number of events by some fixed time t. To do so we would successively simulate the interevent times, keeping track of their sum (which represent the times at which events occur) until that sum exceeds t. That is, we keep on generating interevent times until we reach the first event time after t. Letting $N(t)$—the raw simulation estimator—denote the number of simulated events by time t, we find that a natural quantity to use as a control variable is the sequence of $N(t) + 1$ interevent times that were generated. That is, if we let μ denote the mean interevent time, then as the random variables $T_i - \mu$ have mean 0 it follows that

$$E\left[\sum_{i=1}^{N(t)+1} (T_i - \mu)\right] = 0$$

Hence we can control by using an estimator of the type

$$N(t) + c\left[\sum_{i=1}^{N(t)+1} (T_i - \mu)\right] = N(t) + c\left[\sum_{i=1}^{N(t)+1} T_i - \mu(N(t) + 1)\right]$$

$$= N(t) + c[S_{N(t)+1} - \mu N(t) - \mu]$$

Now since S_n represents the time of the nth event and $N(t) + 1$ represents the number of events by time t plus 1, it follows that $S_{N(t)+1}$ represents the time of the first event after time t. Hence, if we let $Y(t)$ denote the time from t until the next event [$Y(t)$ is commonly called the excess life at t], then

$$S_{N(t)+1} = t + Y(t)$$

and so the above controlled estimator can be written

$$N(t) + c[t + Y(t) - \mu N(t) - \mu]$$

The best c is given by

$$c^* = -\frac{\text{Cov}[N(t),\ Y(t) - \mu N(t)]}{\text{Var}[Y(t) - \mu N(t)]}$$

Now for t large, it can be shown that the terms involving $N(t)$ dominate—because their variance will grow linearly with t, whereas the other terms will remain bounded—and so for t large

$$c^* \approx -\frac{\text{Cov}[N(t), -\mu N(t)]}{\text{Var}[-\mu N(t)]} = \frac{\mu \text{ Var}[N(t)]}{\mu^2 \text{ Var}[N(t)]} = \frac{1}{\mu}$$

Thus for t large, the best controlled estimator of the above type is close to

$$N(t) + \frac{1}{\mu}(t + Y(t) - \mu N(t) - \mu) = \frac{Y(t)}{\mu} + \frac{t}{\mu} - 1 \tag{8.7}$$

In other words, for t large, the critical value to be determined from the simulation is $Y(t)$, the time from t until the next renewal.

The above estimator can further be improved upon by the use of "conditioning." Namely, rather than using the actual observed time of the first event after t, we can condition on $A(t)$, the time at t since the last event (see Figure 8.3). The quantity $A(t)$ is often called the age of the renewal process at t. [If we imagine a system consisting of a single item that functions for a random time having distribution F and then fails and is immediately replaced by a new item, then we have a renewal process with each event corresponding to the failure of an item. The variable $A(t)$ would then refer to the age of the item in use at time t, where by age we mean the amount of time it has already been in use.]

Now if the age of the process at time t is x, the expected remaining life of the item is just the expected amount by which an interevent time exceeds x given that it is greater than x. That is,

$$\begin{aligned} E[Y(t) \mid A(t) = x] &= E[T - x \mid T > x] \\ &= \int_x^\infty (y - x) \frac{f(x)\,dx}{1 - F(x)} \\ &\equiv \mu[x] \end{aligned}$$

where the above supposes that F is a continuous distribution with density function f. Hence, with $\mu[x]$ defined as above to equal $E[T - x \mid T > x]$, we see that

$$E[Y(t) \mid A(t)] = \mu[A(t)]$$

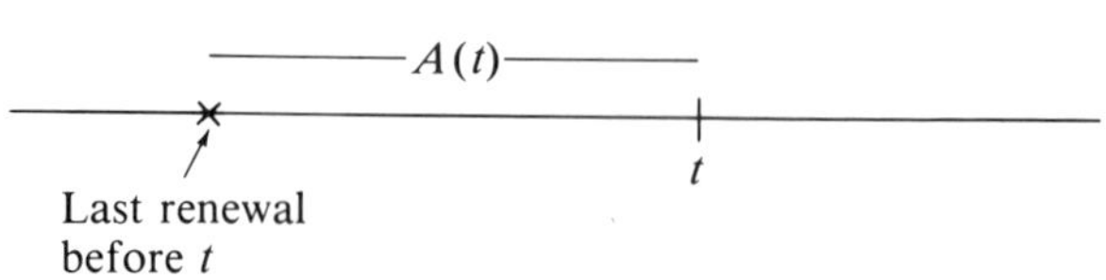

Figure 8.3 Age at t.

Thus for large t, a better estimator of $E[N(t)]$ than the one given in Equation (8.7) is

$$\frac{\mu[A(t)]}{\mu} + \frac{t}{\mu} - 1 \tag{8.8}$$

8.4. Stratified Sampling

Consider a finite capacity queueing system which has the property that it empties at the end of each day and then begins anew at the beginning of the next day. Customers arrive daily in accordance with a Poisson process, but suppose that the rate of that process can change each day. Specifically, suppose that, for any given day, the Poisson arrival rate, independently of what occurred on prior days, is either λ_1, λ_2, or λ_3, each with probability $\frac{1}{3}$. We are interested in simulating this system so as to estimate the average number of customers that are lost (because they arrived when the system was at capacity) during a day.

To simulate the above we begin each run by generating a random variable I which is equally likely to be either 1, 2, or 3. We then simulate one day of the above system under the assumption that customers arrive in accordance with a Poisson process having rate λ_I. Intuitively, however, since we know that each day is equally likely to have Poisson arrivals at rate λ_i, $i = 1, 2, 3$, it seems that it might be preferable to perform exactly one-third of our simulation runs using each of these three rates.

To verify the above intuition, let, for $i = 1, 2, 3$, L_i^* denote the number of lost customers in a day given that the arrivals during that day are Poisson with mean λ_i. Also, let L denote the number of lost customers in a day. Then, upon conditioning upon what the daily arrival rate is, we see that

$$E[L] = \tfrac{1}{3}\,(E[L_1^*] + E[L_2^*] + E[L_3^*]) \tag{8.9}$$

Now let L_i, having the same distribution as L, denote the number of lost customers on day i. We now argue that

$$\mathrm{Var}(L_1^* + L_2^* + L_3^*) \le \mathrm{Var}(L_1 + L_2 + L_3) \tag{8.10}$$

Since $(L_1^* + L_2^* + L_3^*)/3$ is by (8.9) an unbiased estimator of $E[L]$, it thus follows from (8.10) that using each λ_i in exactly one-third of the simulation runs (as opposed to a random choosing at the beginning of each run) results in an unbiased estimator having smaller variance.

To prove the inequality (8.10) we make use of the conditional variance formula

$$\mathrm{Var}(X) = E[\mathrm{Var}(X|Y)] + \mathrm{Var}(E[X|Y])$$

Because both terms on the right are nonnegative, we see that

$$\text{Var}(X) \geq E[\text{Var}(X|Y)]$$

Hence, with L representing the number of lost customers in a day and I being the index of the Poisson arrival rate for that day, we see that

$$\begin{aligned}\text{Var}(L) &\geq E[\text{Var}(L|I)] \\ &= \sum_{i=1}^{3} \text{Var}(L|I = i)\, P\{I = i\} \\ &= \tfrac{1}{3} \sum_{i=1}^{3} \text{Var}(L_i^*) \\ &= \tfrac{1}{3} \text{Var}(L_1^* + L_2^* + L_3^*)\end{aligned}$$

which, since $\text{Var}(L_1 + L_2 + L_3) = 3\ \text{Var}(L)$ establishes (8.10).

The above approach, which does not randomly select the arrival rate for a simulation run but rather lets the proportion of runs with each arrival rate be fixed according to the probability that an arbitrary run has that arrival rate, is called *stratified sampling*.

REMARKS

1. It is interesting that the proof that stratified sampling leads to a reduction in variance uses the conditional variance formula to assert that $\text{Var}(X) \geq E[\text{Var}(X|Y)]$, whereas the proof that "conditioning" always reduces the variance of an estimator (see Section 8.3) uses the conditional variance formula to assert that $\text{Var}(X) \geq \text{Var}(E[X|Y])$.

2. The above shows that performing exactly one-third of the simulations with each of the three arrival rates is better than randomly choosing one of these rates at the beginning of each run (with each rate being equally likely to be chosen). However, this does not solve the problem of determining the optimal number of runs to be performed for each arrival rate. To consider this problem suppose that we want to perform a total of n simulation runs. If we now performed n_i of those runs using arrival rate λ_i, $i = 1, 2, 3$, where $n_1 + n_2 + n_3 = n$, then the estimator of $E[L]$ would be $(\bar{L}_1^* + \bar{L}_2^* + \bar{L}_3^*)/3$, where $\bar{L}_i^*$ is the average number lost daily for the n_i runs using arrival rate λ_i. The variance of this estimator is thus one-ninth of $\sigma_1^2/n_1 + \sigma_2^2/n_2 + \sigma_3^2/n_3$, where σ_i^2 is the variance of the number of lost customers in a day given that the arrival rate for that day is λ_i. Whereas the values σ_i^2 would certainly be initially unknown, we could perform a small simulation to estimate them—say, we use s_i^2 to estimate σ_i^2. We could now choose n_1, n_2, n_3 by solving the following optimization problem:

$$\text{Minimize} \left\{ \sum_{i=1}^{3} \frac{s_i^2}{n_i} : n_1 + n_2 + n_3 = n \right.$$ ■

For another application of stratified sampling suppose we wanted to estimate

$$\theta = \int_0^1 h(x)\, dx = E[h(U)]$$

Rather than generating n random numbers $U_1, \ldots, U_n$ and using $\Sigma_{i=1}^n h(U_i)/n$ as our estimator of θ, the stratified sampling approach would break up the interval (0,1) into the n equal parts (0, 1/n), (1/n, 2/n), . . . , ((n − 1)/n, 1) and then generate a random variable uniformly distributed in each of these intervals. That is, let $\overline{U}_i$ be uniform in $((i-1)/n, i/n)$, $i = 1, \ldots, n$; to generate such a random variable set $\overline{U}_i = (i - 1 + U_i)/n$, where U_i is a random number. The stratified sampling approach would then estimate θ by $\Sigma_{i=1}^n h(\overline{U}_i)/n$. In the same manner as was shown in the beginning of this section, we can show that this stratified estimator remains unbiased and has a smaller variance (see Problem 24) than the estimator $\Sigma_{i=1}^n (U_i)/n$.

For another use of stratified sampling suppose we wanted to use simulation to estimate

$$\theta = E[h(Y_1, \ldots, Y_m)]$$

where the Y_i are independent, with Y_i having distribution function F_i. Suppose also that the value of y_1 has a significantly large effect on the value of $h(y_1, \ldots, y_m)$. If we are planning to generate n simulation runs—where each run consists of generating $Y_1, \ldots, Y_m$ and computing $h(Y_1, \ldots, Y_m)$—and we are going to use the inverse transform method to generate the successive values of Y_1, then a more effective procedure than generating the n values of Y_1 by using $F_1^{-1}(U_i)$, $i = 1, \ldots, n$, where U_i are independent random numbers, is to let the first random number used be uniform over $(0, 1/n)$, let the second be uniform over $(1/n, 2/n)$, and so on. That is, the value of Y_1 used in the ith simulation run should be obtained from $F_1^{-1}((i - 1 + U_i)/n)$.

8.5. Importance Sampling

Let $\mathbf{Y} = (Y_1, \ldots, Y_m)$ denote a vector of random variables having a joint probability density function $f(y_1, \ldots, y_m)$, and suppose that we are interested in estimating

$$\theta = E[h(Y_1, \ldots, Y_m)] = \int h(y_1, \ldots, y_m) f(y_1, \ldots, y_m)\, dy_1 \cdots dy_m$$

where the above is meant to be an m-dimensional integral. (If the Y_i are discrete, interpret f as the joint mass function and regard the integral as an m-fold summation.)

Suppose that a direct simulation of the random vector $\mathbf{Y}$ so as to compute values of $h(\mathbf{Y})$ is inefficient, possibly because (a) it is difficult to simulate the random vector $\mathbf{Y}$, or (b) the variance of $h(\mathbf{Y})$ is large, or (c) a combination of (a) and (b).

Another way in which we can use simulation to estimate θ is to note that for any other random vector $\mathbf{W} = (W_1, \ldots, W_m)$, which takes values in the same region as does $\mathbf{Y}$ and has joint density function $g(w_1, \ldots, w_m)$, θ can be expressed as

$$\begin{aligned}\theta &= \int \frac{h(w_1, \ldots, w_m) f(w_1, \ldots, w_m)}{g(w_1, \ldots, w_m)} g(w_1, \ldots, w_m)\, dw_1 \cdots dw_m \\ &= E_g[h(W_1, \ldots, W_m) f(W_1, \ldots, W_m)/g(W_1, \ldots, W_m)] \qquad (8.11)\end{aligned}$$

where we have written E_g to emphasize that the random vector $\mathbf{W}$ has joint density g.

It follows from Equation (8.11) that we can estimate θ by successively generating values of the random vector $\mathbf{W}$ and then using as the estimator the resulting average of the values of $h(\mathbf{W})f(\mathbf{W})/g(\mathbf{W})$. If the density function $g(\mathbf{w})$ can be chosen to be similar in shape to the function $h(\mathbf{w})f(\mathbf{w})$, then their ratio will not vary much and so $h(\mathbf{W})f(\mathbf{W})/g(\mathbf{W})$ will have a small variance. Unfortunately, however, this is often difficult to accomplish and usually a small simulation experiment is necessary to see if this approach—referred to as importance sampling—does indeed result in an acceptably small variance.

REMARK

The importance sampling approach is also useful in that it enables us to estimate two (or more) distinct quantities in a single simulation. For example, suppose that

$$\theta_1 = E[h(\mathbf{Y})] \qquad \text{and} \qquad \theta_2 = E[h(\mathbf{W})]$$

where $\mathbf{Y}$ and $\mathbf{W}$ are random vectors having joint density functions f and g, respectively. If we now simulate $\mathbf{W}$, we can simultaneously use $h(\mathbf{W})$ and $h(\mathbf{W})f(\mathbf{W})/g(\mathbf{W})$ as estimators of θ_2 and θ_1, respectively. For example, suppose we simulate T, the total time in the system of the first r customers in a queueing system in which the service distribution is exponential with mean 2. If we now decide that we really should have considered the same system but with a service distribution that is gamma distributed with parameters (2, 1), then it is not necessary to repeat the simulation, because we can just use the estimator

$$T\frac{\prod_{i=1}^{r} S_i\exp\{-S_i\}}{\prod_{i=1}^{r} (\frac{1}{2}\exp\{-S_i/2\})} = 2^r T \exp\left\{-\sum_{i=1}^{r}\frac{S_i}{2}\right\}\prod_{i=1}^{r} S_i$$

where S_i is the (exponentially) generated service time of customer i. [The above follows since the exponential service time density is $g(s) = \frac{1}{2}e^{-s/2}$ whereas the gamma (2, 1) density is $f(s) = se^{-s}$.]

Importance sampling is very powerful when estimating small probabilities ■

Example 8m: Suppose we want to compute the probability that a particular n-component system having structure function ϕ (see Example 8b) will function when the components act independently and component i functions with probability p_i, $i = 1, \ldots, n$. If S_i is the indicator variable for component i—being 1 if i functions and 0 if i is failed—then

$$P\{S_i = s\} = p_i^s(1 - p_i)^{1-s}, \qquad s = 0, 1$$

In cases where the probability that the system functions is very near 1, it would require a large number of simulations of the vector $(S_1, \ldots, S_n)$ to obtain an estimate that is highly accurate. (See Problem 28, where it is shown that if the probability of failure is around 10^{-6} and one wanted their estimate to be, with roughly 95 percent confidence, correct to within $\pm 10^{-6}$, then it would require roughly 4 million simulation runs.) One possibility for reducing the large number of simulation runs that would be needed would be to simulate the system according to a different set of component probabilities—say, q_i, $i = 1, \ldots, n$. If we now simulate S_i, $i = 1, \ldots, n$, according to these q_i and let

$$I = 1 - \phi = \begin{cases} 1 & \text{if the resultant system fails} \\ 0 & \text{otherwise} \end{cases}$$

then the quantity

$$\frac{I\prod_{i=1}^{n} (p_i)^{S_i}(1 - p_i)^{1-S_i}}{\prod_{i=1}^{n} (q_i)^{S_i}(1 - q_i)^{1-S_i}}$$

can be used to estimate the probability that the system fails under the component reliabilities p_i, $i = 1, \ldots, n$.

For example, consider a 50 of 100 system (i.e., the system will function if at least 50 of the 100 components function), where component i functions with probability p_i, $i = 1, \ldots, 100$, and where all the p_i are greater than .7. Such a system is highly unlikely to fail and so it may be more efficient to simulate a 50 of 100 system having component reliabilities $q_i \equiv \frac{1}{2}$. If S_i, equally likely to be 0 or 1, is the generated value of component i for this system, the failure probability can be estimated by

$$I 2^n \prod_{i=1}^{n} (p_i)^{S_i} (1 - p_i)^{1-S_i}$$

where I is 1 if $\Sigma_{i=1}^{100} S_i < 50$. (When all p_i equals .7, the variance of the above estimator is 3×10^{-9} as compared to 2×10^{-5} for the estimator I.) ■

Example 8n: Let $X_1, X_2, \ldots$ be a sequence of independent and identically distributed normal random variables having mean μ and variance 1, where $\mu < 0$. An important problem in the theory of quality control (specifically in the analysis of the so-called cumulative sum charts) is to determine the probability that the partial sums of these values exceeds B before going below $-A$. That is, let

$$S_n = \sum_{i=1}^{n} X_i$$

and define

$$N = \text{Min}\{n\text{: either } S_n < -A, \text{ or } S_n > B\}$$

where A and B are fixed positive numbers. We are now interested in estimating

$$\theta = P\{S_N > B\}$$

An effective way of estimating θ is by simulating the X_i as if they were normal with mean $-\mu$ and variance 1, stopping again when their sum either exceeds B or falls below $-A$. (Since $-\mu$ is positive, the stopped sum is greater than B more often than if we were simulating with the original negative mean.) If $X_1, \ldots, X_N$ denote the simulated variables (each being normal with mean $-\mu$ and variance 1) and

$$I = \begin{cases} 1 & \text{if } \sum_{i=1}^{N} X_i > B \\ 0 & \text{otherwise} \end{cases}$$

then the estimate of θ from this run is

$$I \prod_{i=1}^{N} \left[\frac{f_\mu(X_i)}{f_{-\mu}(X_i)} \right] \tag{8.12}$$

where f_c is the normal density with mean c and variance 1. Since

$$\frac{f_\mu(x)}{f_{-\mu}(x)} = \frac{\exp\left\{-\dfrac{(x-\mu)^2}{2}\right\}}{\exp\left\{-\dfrac{(x+\mu)^2}{2}\right\}} = e^{2\mu x}$$

it follows from (8.12) that the estimator of θ based on this run is

$$I \exp\left\{2\mu \sum_{i=1}^{N} X_i\right\} = I \exp\{2\mu S_N\}$$

When I is equal to 1, S_N exceeds B and, since $\mu < 0$, the estimator in this case is less than $e^{2\mu B}$. That is, rather than obtaining from each run either the value 0 or 1—as would occur if we did a straight simulation—we obtain in this case either the value 0 or a value that is less than $e^{2\mu B}$; which strongly indicates why this importance sampling approach results in a reduced variance. For example, if $\mu = -0.1$ and $B = 5$, then the estimate from each run lies between 0 and $e^{-1} = 0.3679$. In addition, the above is theoretically important because it shows that

$$P\{\text{cross } B \text{ before } -A\} \leq e^{-2\mu B}$$

Since the above is true for all positive A, we obtain the interesting result

$$P\{\text{cross } B \text{ before going negative}\} \leq e^{2\mu B} \qquad \blacksquare$$

8.6. Using Common Random Numbers

Suppose that a set of n jobs are to be processed by either of a pair of identical machines. Let T_i denote the processing time for job i, $i = 1, \ldots, n$. We are interested in comparing the time it takes to complete the processing of all the jobs under two different policies for deciding the order in which to process jobs. Whenever a machine becomes free, the first policy, called longest job first, always chooses the remaining job having the longest processing time; whereas the second policy, called shortest job first, always selects the one having the shortest processing time. For example, if $n = 3$ and $T_1 = 2$, $T_2 = 5$, and $T_3 = 3$, then the longest job first would complete processing at time 5, whereas the shortest job first would not get done until time 7. We would like to use simulation to compare the expected difference in the completion times under these two policies when the times to process jobs, $T_1, \ldots, T_n$, are random variables having a given distribution F.

In other words, if $g(t_1, \ldots, t_n)$ is the time it takes to process the n jobs having processing times $t_1, \ldots, t_n$ when we use the longest job first policy and if $h(t_1, \ldots, t_n)$ is the time when we use the shortest first policy, then we are interested in using simulation to estimate

$$\theta = \theta_1 - \theta_2$$

where

$$\theta_1 = E[g(\mathbf{T})], \qquad \theta_2 = E[h(\mathbf{T})], \qquad \mathbf{T} = (T_1, \ldots, T_n)$$

If we now generate the vector $\mathbf{T}$ to compute $g(\mathbf{T})$, the question arises whether we should use those same generated values to compute $h(\mathbf{T})$ or whether it is more efficient to generate independent sets to estimate θ_2. To answer this question suppose that we used $\mathbf{T}^* = (T_1^*, \ldots, T_n^*)$, having the same distribution as $\mathbf{T}$, to estimate θ_2. Then the variance of the estimator $g(\mathbf{T}) - h(\mathbf{T}^*)$ of θ is

$$\begin{aligned}\text{Var}(g(\mathbf{T}) - h(\mathbf{T}^*)) &= \text{Var}(g(\mathbf{T})) + \text{Var}(h(\mathbf{T}^*)) - 2\,\text{Cov}(g(\mathbf{T}), h(\mathbf{T}^*)) \\ &= \text{Var}(g(\mathbf{T})) + \text{Var}(h(\mathbf{T})) - 2\,\text{Cov}(g(\mathbf{T}), h(\mathbf{T}^*)) \quad (8.13)\end{aligned}$$

Hence, if $g(\mathbf{T})$ and $h(\mathbf{T})$ are positively correlated—that is, if their covariance is positive—then the variance of the estimator of θ is smaller if we use the same set of generated random values $\mathbf{T}$ to compute both $g(\mathbf{T})$ and $h(\mathbf{T})$ than it would be if we used an independent set $\mathbf{T}^*$ to compute $h(\mathbf{T}^*)$ [in this latter case the covariance in (8.13) would be 0].

Since both g and h are increasing functions of their arguments, it follows, since increasing functions of independent random variables are positively correlated (see the Appendix of this chapter for a proof), that in the above case it is more efficient to successively compare the policies by always using the same set of generated job times for both policies.

As a general rule of thumb when comparing different operating policies in a randomly determined environment, after the environmental state has been simulated one should then evaluate all the policies for this environment. That is, if the environment is determined by the vector $\mathbf{T}$ and $g_i(\mathbf{T})$ is the return from policy i under the environmental state $\mathbf{T}$, then after simulating the value of the random vector $\mathbf{T}$ one should then evaluate, for that value of $\mathbf{T}$, all the returns $g_i(\mathbf{T})$.

Appendix: Proof That Antithetic Variables Lead to a Variance Reduction When $h(x_1, \ldots, x_n)$ Is Monotone in Each of Its Coordinates

The following theorem is the key to showing that the use of antithetic variables will lead to a reduction in variance in comparison with generating a new independent set of random numbers whenever the function h is monotone in each of its coordinates.

Theorem

If $X_1, \ldots, X_n$ *are independent, then for any increasing functions* f *and* g *of* n *variables*

$$E[f(\mathbf{X})g(\mathbf{X})] \geq E[f(\mathbf{X})]E[g(\mathbf{X})] \tag{1}$$

where $\mathbf{X} = (X_1, \ldots, X_n)$.

Proof.† The proof is by induction on n. To prove it when $n = 1$, let f and g be increasing functions of a single variable. Then for any x and y

$$[f(x) - f(y)][g(x) - g(y)] \geq 0$$

since if $x \geq y$ ($x \leq y$) then both factors are nonnegative (nonpositive). Hence, for any random variables X and Y,

$$[f(X) - f(Y)][g(X) - g(Y)] \geq 0$$

implying that

$$E\{[f(X) - f(Y)][g(X) - g(Y)]\} \geq 0$$

or, equivalently,

$$E[f(X)g(X)] + E[f(Y)g(Y)] \geq E[f(X)g(Y)] + E[f(Y)g(X)]$$

If we now suppose that X and Y are independent and identically distributed then, as in this case,

$$\begin{aligned} E[f(X)g(X)] &= E[f(Y)g(Y)] \\ E[f(X)g(Y)] &= E[f(Y)g(X)] = E[f(X)]E[g(X)] \end{aligned}$$

we obtain the result when $n = 1$.

So assume that Equation (1) holds for $n - 1$ variables, and now suppose that $X_1, \ldots, X_n$ are independent and f and g are increasing functions. Then

$$\begin{aligned} & E[f(\mathbf{X})g(\mathbf{X}) \mid X_n = x_n] \\ &\quad = E[f(X_1, \ldots, X_{n-1}, x_n)g(X_1, \ldots, X_{n-1}, x_n) \mid X_n = x] \\ &\quad = E[f(X_1, \ldots, X_{n-1}, x_n)g(X_1, \ldots, X_{n-1}, x_n)] \\ &\qquad \text{by independence} \\ &\quad \geq E[f(X_1, \ldots, X_{n-1}, x_n)]E[g(X_1, \ldots, X_{n-1}, x_n)] \\ &\qquad \text{by the induction hypothesis} \\ &\quad = E[f(\mathbf{X}) \mid X_n = x_n]E[g(\mathbf{X}) | X_n = x_n] \end{aligned}$$

Hence

$$E[f(\mathbf{X})g(\mathbf{X}) \mid X_n] \geq E[f(\mathbf{X}) \mid X_n]E[g(\mathbf{X}) | X_n]$$

† From S. Ross, *Introduction to Probability Models*, 4th Ed., New York: Academic Press, 1989, with permission.

and, upon taking expectations of both sides,

$$E[f(\mathbf{X})g(\mathbf{X})] \geq E[E[f(\mathbf{X})|X_n]E[g(\mathbf{X}) \mid X_n]]$$
$$\geq E[f(\mathbf{X})]E[g(\mathbf{X})]$$

The last inequality follows because $E[f(\mathbf{X}) \mid X_n]$ and $E[g(\mathbf{X}) \mid X_n]$ are both increasing functions of X_n, and so, by the result for $n = 1$,

$$E[E[f(\mathbf{X}) \mid X_n]E[g(\mathbf{X}) \mid X_n]] \geq E[E[f(\mathbf{X}) \mid X_n]]E[E[g(\mathbf{X}) \mid X_n]]$$
$$= E[f(\mathbf{X})]E[g(\mathbf{X})] \qquad \square$$

Corollary:

If $h(x_1, \ldots, x_n)$ is a monotone function of each of its arguments, then, for a set $U_1, \ldots, U_n$ of independent random numbers,

$$\text{Cov}[h(U_1, \ldots, U_n), h(1 - U_1, \ldots, 1 - U_n)] \leq 0$$

Proof. By redefining h we can assume, without loss of generality, that h is increasing in its first r arguments and decreasing in its final $n - r$. Hence, letting

$$f(x_1, \ldots, x_n) = h(x_1, \ldots, x_r, 1 - x_{r+1}, \ldots, 1 - x_n)$$
$$g(x_1, \ldots, x_n) = -h(1 - x_1, \ldots, 1 - x_r, x_{r+1}, \ldots, x_n)$$

it follows that f and g are both increasing functions. Thus by the preceding theorem,

$$\text{Cov}[f(U_1, \ldots, U_n), g(U_1, \ldots, U_n)] \geq 0$$

or, equivalently,

$$\text{Cov}[h(U_1, \ldots, U_r, 1 - U_{r+1}, \ldots, 1 - U_n), h(1 - U_1, \ldots, 1 - U_r, U_{r+1}, \ldots, U_n)] \leq 0$$

The result now follows since the random vector $h(U_1, \ldots, U_n), h(1 - U_1, \ldots, 1 - U_n)$ has the same joint distribution as does the random vector

$$h(U_1, \ldots, U_r, 1 - U_{r+1}, \ldots, 1 - U_n),$$
$$h(1 - U_1, \ldots, 1 - U_r, U_{r+1}, \ldots, U_n). \square$$

Problems

1. Suppose we wanted to estimate θ, where

$$\theta = \int_0^1 e^{x^2}\, dx$$

Show that generating a random number U and then using the estimator $e^{U^2}(1 + e^{1-2U})/2$ is better than generating two random numbers U_1 and U_2 and using $[\exp(U_1^2) + \exp(U_2^2)]/2$.

2. Explain how antithetic variables can be used in obtaining a simulation estimate of the quantity

$$\theta = \int_0^1 \int_0^1 e^{(x+y)^2}\, dy\, dx$$

Is it clear in this case that using antithetic variables is more efficient than generating a new pair of random numbers?

3. Let X_i, $i = 1, \ldots, 5$, be independent exponential random variables each with mean 1, and consider the quantity θ defined by

$$\theta = P\left\{\sum_{i=1}^{5} iX_i \geq 21.6\right\}$$

(a) Explain how we can use simulation to estimate θ.
(b) Give the antithetic variable estimator.
(c) Is the use of antithetic variables efficient in this case?

4. Show that if X and Y have the same distribution then $\text{Var}[(X + Y)/2] \leq \text{Var}(X)$; and conclude that the use of antithetic variables can never increase variance (although it need not be as efficient as generating an independent set of random numbers).

5. (a) If Z is a unit normal random variable, design a study using antithetic variables to estimate $\theta = E[Z^3 e^Z]$.
(b) Using the above, do the simulation to obtain an interval of length no greater than 0.1, which you can assert, with 95 percent confidence, contains the value of θ.

6. Verify Equation (8.1).

7. Verify Equation (8.2).

8. Explain why it is unusual to obtain a large increase in efficiency from using both control variables and antithetic variables as opposed to just using one or the other.

9. In certain situations a random variable X, whose mean is known, is simulated so as to obtain an estimate of $P\{X \leq a\}$ for a given constant a. The raw simulation estimator from a single run is I, where

$$I = \begin{cases} 1 & \text{if } X \leq a \\ 0 & \text{if } X > a \end{cases}$$

Because I and X are clearly negatively correlated, a natural attempt to reduce the variance is to use X as a control—and so use an estimator of form $I + c(X - E[X])$.

(a) Determine the percentage of variance reduction over the raw estimator I that is possible (by using the best c) if X were uniform on $(0, 1)$.
(b) Repeat (a) if X were exponential with mean 1.
(c) Explain why we knew that I and X were negatively correlated.

10. Show that $\text{Var}(\alpha X + (1 - \alpha)W)$ is minimized by α being equal to the value given in Eq. 8.4.

11. (a) Explain how control variables may be used to estimate θ in Problem 1.
(b) Do 100 simulation runs, using the control given in (a), to estimate first c^* and then the variance of the estimator.
(c) Using the same data as in part (b), determine the variance of the antithetic variable estimator.
(d) Which of the two types of variance reduction technique worked better in this example.

12. Repeat Problem 11 for θ as given in Problem 2.

13. Repeat Problem 11 for θ as given in Problem 3.

14. Show that in estimating $\theta = E[(1 - U^2)^{1/2}]$ it is better to use U^2 rather than U as the control variate. To do this, use simulation to approximate the necessary covariances.

15. Five elements, numbered 1, 2, 3, 4, 5, are initially arranged in a random order (i.e., the initial ordering is a random permutation of 1, 2, 3, 4, 5). At each stage one of the elements is selected and put at the front of the list. That is, if the present order is 2, 3, 4, 1, 5 and element 1 is chosen, then the new ordering is 1, 2, 3, 4, 5. Suppose that each selection is, independently, element i with probability p_i, where $p_1 = \frac{1}{15}$, $p_2 = \frac{2}{15}$, $p_3 = \frac{3}{15}$, $p_4 = \frac{4}{15}$, $p_5 = \frac{5}{15}$. Let L_j denote the position of the jth element to be selected, and let $L = \sum_{j=1}^{100} L_j$. We are interested in using simulation to estimate $E[L]$.

(a) Explain how we could use simulation to estimate $E[L]$.

(b) Compute $E[N_i]$, where N_i is the number of times element i is chosen in the 100 selections.
(c) Let $Y = \Sigma_{i=1}^{5} iN_i$. Do you think Y is positively or negatively correlated with L?
(d) Develop a study to estimate L, using Y as a control variable.

16. (A Blackjack Project). Suppose we are interested in determining the average winnings (or losses) of an individual, using some given strategy (to be supplied by the user), playing against a dealer. Suppose that a single deck of cards is employed and that the deck is reshuffled after every 5 games. Let a simulation run be the play of these five hands. Assuming that a reshuffle corresponds to a random permutation of the 52 cards in a deck, we are interested in using simulation to estimate the average winnings per run. To improve efficiency we want to use a control variable. To do so, let N_i denote the number of hands in the run for which the player's first two cards sum to i or higher, and compute $E[N_i]$ for $i = 17, 18, 19, 20, 21$. (For example, $E[N_{21}]/5 = 2(\frac{1}{13})(\frac{4}{13})$; $E[N_{20}]/5 = (\frac{4}{13})(\frac{15}{51}) + 2(\frac{1}{13})(\frac{1}{13})$.) Do a simulation study to determine which of the five variables N_i is the most efficient to use as a control variate.

17. Let X and Y be independent with respective distributions F and G and with expected values μ_x and μ_y. For a given value t, we are interested in estimating $\theta = P\{X + Y \leq t\}$.
(a) Give the raw simulation approach to estimating θ.
(b) Use "conditioning" to obtain an improved estimator.
(c) Give a control variable that can be used to further improve upon the estimator in (b).
(d) If F and G are easily inverted, explain how antithetic variables can be used to improve upon the estimator in (b).

18. (The Hit–Miss Method). Let g be a bounded function over the interval [0, 1]—for example, suppose $0 \leq g(x) \leq b$ whenever $0 \leq x \leq 1$—and suppose we are interested in using simulation to approximate $\theta = \int_0^1 g(x)dx$. The hit–miss method for accomplishing this is to generate a pair of independent random numbers U_1 and U_2. Now set $X = U_1$, $Y = bU_2$ so that the random point (X, Y) is uniformly distributed in a rectangle of length 1 and height b. Now set

$$I = \begin{cases} 1 & \text{if } Y < g(x) \\ 0 & \text{otherwise} \end{cases}$$

That is, I is equal to 1 if the random point (X, Y) falls within the shaded area of Figure 8.4.

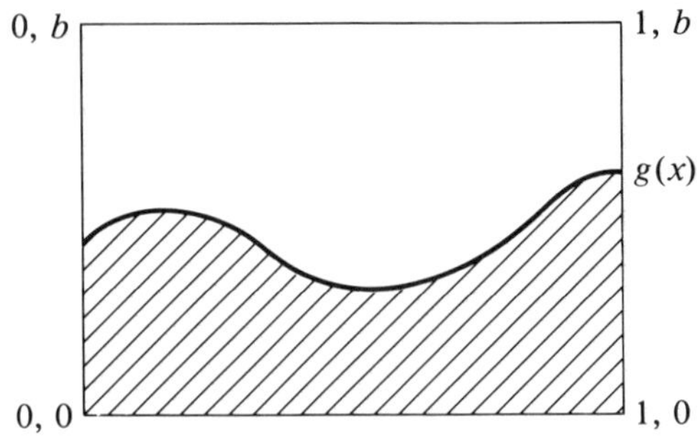

Figure 8.4

(a) Show that $E[I] = [\int_0^1 g(x)\, dx]/b$

(b) Show that $\text{Var}(bI) \geq \text{Var}(g(U))$ and so the hit–miss estimator has a larger variance than simply computing g of a random number.

19. Suppose that X is exponential with mean 1; and given that $X = x$, Y is exponential with mean x. (Thus X and Y are dependent random variables.) Give an efficient way to estimate $P\{XY \leq 3\}$.

20. In Example 81 suppose there are k servers, each of whose service time is exponential with mean μ. Give an efficient estimator of θ.

21. In Example 81 suppose that there is a single server whose service distribution is G. To use the improved estimator of θ, what would S_i represent? What is this estimator?

22. Consider a single server queue where customers arrive according to a Poisson process with rate 2 per minute and the service times are exponentially distributed with mean 1 minute. Let T_i denote the amount of time that customer i spends in the system. We are interested in using simulation to estimate $\theta = E[T_1 + \cdots + T_{10}]$.

(a) Do a simulation to estimate the variance of the raw simulation estimator. That is, estimate $\text{Var}(T_1 + \cdots + T_{10})$.

(b) Do a simulation to determine the improvement over the raw estimator obtained by using antithetic variables.

(c) Do a simulation to determine the improvement over the raw estimator obtained by using $\Sigma_{i=1}^{10} S_i$ as a control variate, where S_i is the ith service time.

(d) Do a simulation to determine the improvement over the raw estimator obtained by using $\Sigma_{i=1}^{10} S_i - \Sigma_{i=1}^{9} I_i$ as a control variate, where I_i is the time between the ith and $(i + 1)$st arrival.

(e) Do a simulation to determine the improvement over the raw estimator obtained by using the estimator $\Sigma_{i=1}^{10} E[T_i \mid N_i]$, where N_i is the number in the system when customer i arrives (and so $N_1 = 0$).

23. (a) A Poisson process having rate λ is equivalent to a renewal process having which interarrival distribution?

(b) Let $N(t)$ denote the number of events by time t of a Poisson process having rate λ. Determine the variance of the estimate of $E[N(t)]$ given by Equation (8.7) and compare it with the variance of the estimator given by (8.8).

Note: The purpose of this problem is not to actually estimate $E[N(t)]$, which we know is just λt, but rather to determine how much better the estimator (8.8) is than the estimator (8.7) in a case where these variances can be explicitly determined.

In the following two problems we are interested in approximating

$$\theta = \int_0^1 h(x)\, dx$$

24. Show that the stratified sampling estimator $(1/n)\Sigma_{i=1}^n h([i - 1 + U_i]/n)$ is a superior estimator than is $(1/n)\Sigma_{i=1}^n h(U_i)$, where the U_i are random numbers.

25. If h is a monotonic function, explain how we can use antithetic variables in conjunction with the stratified sampling estimator in Problem 24 to further improve its efficiency.

26. Explain why, if h is a monotonic function, that $\Sigma_{i=1}^n U_i$ is a natural control variate to use in conjunction with the stratified sampling estimator of Problem 24.

27. Put together the results of Example 8i along with the preceding three problems to obtain an improved simulation estimator of π. Write a program and use this to approximate π.

28. Let X_i, $i \geq 1$, be independent random variables such that $P\{X_i = 1\} = 1 - P\{X_i = 0\} = p$. Show that if one wants to construct a 95 percent confidence interval of p which is of length 2×10^{-6} then we would need a sample size of approximately 4×10^6 when $p = 10^{-6}$.

29. Suppose we have a "black box" which on command can generate the value of a gamma random variable with parameters $\frac{3}{2}$ and 1. Explain how we can use this black box to approximate $E[e^X/(X + 1)]$, where X is an exponential random variable with mean 1.

30. Consider two different approaches for manufacturing a product. The profit from these approaches depends on the value of a parameter α, and let $v_i(\alpha)$ denote the profit of approach i as a function of α. Suppose that approach 1 works best for small values of α in that $v_1(\alpha)$ is a decreasing function of α, whereas approach 2 works best for large values of α in that $v_2(\alpha)$ is an increasing function of α. If the daily value of α is a random variable coming from the distribution F, then in comparing the average profit of these two approaches should we generate a single value of α and compute the profits for this α, or should we generate α_1 and α_2 and then compute $v_i(\alpha_i)$, $i = 1, 2$?

CHAPTER 9

Statistical Validation Techniques

Introduction

In this chapter we consider some statistical procedures that are useful in validating simulation models. Sections 9.1 and 9.2 consider goodness of fit tests, which are useful in ascertaining whether an assumed probability distribution is consistent with a given set of data. In Section 9.1 we suppose that the assumed distribution is totally specified, whereas in Section 9.2 we suppose that it is only specified up to certain parameters—for example, it may be Poisson having an unknown mean. In Section 9.3 we show how one can test the hypothesis that two separate samples of data come from the same underlying population—as would be the case with real and simulated data when the assumed mathematical model being simulated is an accurate representation of reality. The results of Section 9.3 are particularly useful in testing the validity of a simulation model. A generalization to the case of many samples is also presented in this section. Finally, in Section 9.4, we show how to use real data to test the hypothesis that the process generating the data constitutes a nonhomogeneous Poisson process. The case of a homogeneous Poisson process is also considered in this section.

9.1. Goodness of Fit Tests

One often begins a probabilistic analysis of a given phenomena by hypothesizing that certain of its random elements have a particular probability distribution. For example, we might begin an analysis of a traffic network by supposing that the daily number of accidents has a Poisson distribution. Such hypotheses can be statistically tested by observing data and then seeing whether the assumption of a particular probability distribution is consistent with this data. These statistical tests are called *goodness of fit* tests.

One way of performing a goodness of fit test is to first partition the possible values of a random quantity into a finite number of regions. A sample of values of this quantity is then observed and a comparison is made between the numbers of them that fall into each of the regions and the theoretical expected numbers when the specified probability distribution is indeed governing the data.

In this section we consider goodness of fit tests when all the parameters of the hypothesized distribution are specified; and in the following section we consider such tests when certain of the parameters are unspecified. We first consider the case of a discrete and then a continuous hypothesized distribution.

The Chi-Square Goodness of Fit Test for Discrete Data

Suppose that n independent random variables—$Y_1, \ldots, Y_n$—each taking on one of the values $1, 2, \ldots, k$, are to be observed, and that we are interested in testing the hypothesis that $\{p_i, i = 1, \ldots, k\}$ is the probability mass function of these random variables. That is, if Y represents any of the Y_j, the hypothesis to be tested, which we denote by H_0 and refer to as the *null hypothesis*, is

$$H_0\colon P\{Y = i\} = p_i, \qquad i = 1, \ldots, k$$

To test the foregoing hypothesis, let N_i, $i = 1, \ldots, k$, denote the number of the Y_j's that equal i. Because each Y_j independently equals i with probability $P\{Y = i\}$, it follows that, under H_0, N_i is binomial with parameters n and p_i. Hence, when H_0 is true,

$$E[N_i] = np_i$$

and so $(N_i - np_i)^2$ is an indication as to how likely it appears that p_i indeed equals the probability that $Y = i$. When this is large, say, in relation to np_i, then it is an indication that H_0 is not correct. Indeed, such reasoning leads us to consider the quantity

$$T = \sum_{i=1}^{k} \frac{(N_i - np_i)^2}{np_i}$$

and to reject the null hypothesis when T is large.

Whereas small values of the test quantity T are evidence in favor of the hypothesis H_0, large ones are indicative of its falsity. Suppose now that the actual data result in the test quantity T taking on the value t. To see how unlikely such a large outcome would have been if the null hypothesis had been true, we define the so-called p-value by

$$p\text{-value} = P_{H_0}\{T \geq t\}$$

where we have used the notation P_{H_0} to indicate that the probability is to be computed under the assumption that H_0 is correct. Hence the p-value gives the probability that such a large value of T as the one observed would have occurred if the null hypothesis were true. It is typical to reject the null hypothesis—saying that it appears to be inconsistent with the data—when a small p-value results (a value less than .05, or more conservatively, .01 is usually taken to be critical) and to accept the null hypothesis—saying that it appears to be consistent with the data—otherwise.

After observing the value, call it t, of the test quantity, it thus remains to determine the probability

$$p\text{-value} = P_{H_0}\{T \geq t\}$$

A reasonably good approximation to this probability can be obtained by using the classical result that, for large values of n, T has approximately a chi-square distribution with $k - 1$ degrees of freedom when H_0 is true. Hence

$$p\text{-value} \approx P\{X_{k-1}^2 \geq t\} \tag{9.1}$$

where X_{k-1}^2 is a chi-square random variable with $k - 1$ degrees of freedom. The chi-square probability can then be obtained by running Program 9-1.

Example 9a Consider a random quantity which can take on any of the possible values 1, 2, 3, 4, 5 and suppose we want to test the hypothesis that these values are equally likely to occur. That is, we want to test

$$H_0\colon\ p_i = .2, \qquad i = 1, \ldots, 5$$

If a sample of size 50 yielded the following values of N_i:

$$12,\ 5,\ 19,\ 7,\ 7$$

then the approximate p-value is obtained as follows. The value of the test statistic T is given by

$$T = \frac{(4 + 25 + 81 + 9 + 9)}{10} = 12.8$$

Running Program 9-1 yields that

$$p\text{-value} \approx P\{X_4^2 > 12.8\} = .0122$$

For such a low p-value the hypothesis that all outcomes are equally likely would be rejected. ■

If the p-value approximation given by Equation (9.1) is not too small—say, of the order of .15 or larger—then it is clear that the null hypothesis is not going to be rejected and so there is no need to look for a better approximation. However, when the p-value is closer to a critical value (like .05 or .01) we would probably want a more accurate estimate of its value than the one given by the chi-square approximate distribution. Fortunately, a more accurate estimator can be obtained via a simulation study.

The simulation approach to estimating the p-value of the outcome $T = t$ is as follows. To determine the probability that T would have been at least as large as t when H_0 is true, we generate n independent random variables $Y_1^{(1)}, \ldots, Y_n^{(1)}$, each having the probability mass function $\{p_i, i = 1, \ldots, k\}$—that is,

$$P\{Y_j^{(1)} = i\} = p_i, \qquad i = 1, \ldots, k, \quad j = 1, \ldots, n$$

Now let

$$N_i^{(1)} = \text{number } j\text{: } Y_j^{(1)} = i$$

and set

$$T^{(1)} = \sum_{i=1}^{k} \frac{(N_i^{(1)} - np_i)^2}{np_i}$$

Now repeat this procedure by simulating a second set, independent of the first set, of n independent random variables $Y_1^{(2)}, \ldots, Y_n^{(2)}$ each having the probability mass function $\{p_i, i = 1, \ldots, k\}$ and then, as for the first set, determining $T^{(2)}$. Repeating this a large number of times, say, r, yields r independent random variables $T^{(1)}, T^{(2)}, \ldots, T^{(r)}$, each of which has the same distribution as does the test statistic T when H_0 is true. Hence, by the law of large numbers, the proportion of the T_i that are as large as t will be very nearly equal to the probability that T is as large as t when H_0 is true—that is,

$$\frac{\text{number } l: T^{(l)} \geq t}{r} \approx P_{H_0}\{T \geq t\}$$

Program 9-2 uses simulation to approximate the p-value. The user must input the probabilities p_i, $i = 1, \ldots, k$, the sample size n, the value of the observed test quantity T, and the number of simulation runs desired.

Example 9b Let us reconsider Example 9a and this time employ simulation to approximate the p-value. To do so we run Program 9-2.

```
RUN
THIS PROGRAM USES SIMULATION TO APPROXIMATE THE p-value IN THE
   GOODNESS OF FIT TEST
Random number seed (-32768 to 32767)? 6867
ENTER THE NUMBER OF POSSIBLE VALUES
? 5
ENTER THE PROBABILITIES ONE AT A TIME
? .2
```

```
? .2
? .2
? .2
? .2
ENTER THE SAMPLE SIZE
? 50
ENTER THE DESIRED NUMBER OF SIMULATION RUNS
? 1000
ENTER THE VALUE OF THE TEST STATISTIC
? 12.8
THE ESTIMATE OF THE p-value IS .011
OK
```

■

The Kolmogorov–Smirnov Test for Continuous Data

Now consider the situation where $Y_1, \ldots, Y_n$ are independent random variables, and we are interested in testing the null hypothesis H_0 that they have the common distribution function F, where F is a given continuous distribution function. One approach to testing H_0 is to break up the set of possible values of the Y_j into k distinct intervals, say,

$$(y_0, y_1), (y_1, y_2), \ldots, (y_{k-1}, y_k), \qquad \text{where } y_0 = -\infty, \quad y_k = +\infty$$

and then consider the discretized random variables Y_j^d, $j = 1, \ldots, n$, defined by

$$Y_j^d = i \qquad \text{if } Y_j \text{ lies in the interval } (y_{i-1}, y_i)$$

The null hypothesis then implies that

$$P\{Y_j^d = i\} = F(y_i) - F(y_{i-1}), \qquad i = 1, \ldots, k$$

and this can be tested by the chi-square goodness of fit test already presented.

There is, however, another way of testing that the Y_j come from the continuous distribution function F which is generally more efficient than discretizing; and it works as follows. After observing $Y_1, \ldots, Y_n$, let F_e be the empirical distribution function defined by

$$F_e(x) = \frac{\#i\colon\ Y_i \le x}{n}$$

That is, $F_e(x)$ is the proportion of the observed values that are less than or equal to x. Because $F_e(x)$ is a natural estimator of the probability that an observation is less than or equal to x, it follows that, if the null hypothesis that F is the underlying distribution is correct, it should be close to $F(x)$. Since this is so for all x, a natural quantity on which to base a test of H_0 is the test quantity

$$D \equiv \underset{x}{\text{Maximum}}\ |F_e(x) - F(x)|$$

where the maximum (the more mathematically sophisticated readers will recognize that technically we should have written supremum rather than maximum) is over all values of x from $-\infty$ to $+\infty$. The quantity D is called the *Kolmogorov–Smirnov test statistic*.

To compute the value of D for a given data set $Y_j = y_j, j = 1, \ldots, n$, let $y_{(1)}, y_{(2)}, \ldots, y_{(n)}$ denote the values of the y_j in increasing order. That is,

$$y_{(j)} = j\text{th smallest of } y_1, \ldots, y_n$$

For example, if $n = 3$ and $y_1 = 3$, $y_2 = 5$, $y_3 = 1$, then $y_{(1)} = 1$, $y_{(2)} = 3$, $y_{(3)} = 5$. Since $F_e(x)$ can be written

$$F_e(x) = \begin{cases} 0 & \text{if } x < y_{(1)} \\ \dfrac{1}{n} & \text{if } y_{(1)} \le x < y_{(2)} \\ \vdots & \\ \dfrac{j}{n} & \text{if } y_{(j)} \le x < y_{(j+1)} \\ \vdots & \\ 1 & \text{if } y_{(n)} \le x \end{cases}$$

we see that $F_e(x)$ is constant within the intervals $(y_{(j-1)}, y_{(j)})$ and then jumps by $1/n$ at the points $y_{(1)}, \ldots, y_{(n)}$. Since $F(x)$ is an increasing function of x which is bounded by 1, it follows that the maximum value of $F_e(x) - F(x)$ is nonnegative and occurs at one of the points $y_{(j)}, j = 1, \ldots, n$ (see Figure 9.1). That is,

$$\underset{x}{\text{Maximum}} \{F_e(x) - F(x)\} = \underset{j=1,\ldots,n}{\text{Maximum}} \left\{\frac{j}{n} - F(y_{(j)})\right\} \tag{9.2}$$

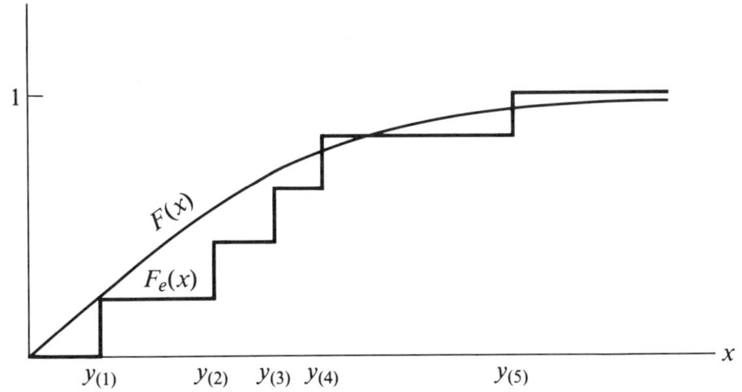

Figure 9.1 $n = 5$.

Similarly, the maximum (supremum) value of $F(x) - F_e(x)$ is also nonnegative and occurs immediately before one of the jump points $y_{(j)}$; and so

$$\underset{x}{\text{Maximum}}\ \{F(x) - F_e(x)\} = \underset{j=1,\ldots,n}{\text{Maximum}} \left\{F(y_{(j)}) - \frac{(j-1)}{n}\right\} \tag{9.3}$$

From Equations (9.2) and (9.3) we see that

$$\begin{aligned} D &= \underset{x}{\text{Maximum}}\ |F_e(x) - F(x)| \\ &= \text{Maximum}\ \{\text{Maximum}\{F_e(x) - F(x)\},\ \text{Maximum}\{F(x) - F_e(x)\}\} \\ &= \text{Maximum}\left\{\frac{j}{n} - F(y_{(j)}),\ F(y_{(j)}) - \frac{(j-1)}{n},\quad j = 1, \ldots, n\right\} \end{aligned} \tag{9.4}$$

Equation (9.4) can be used to compute the value of D.

Suppose now that the Y_j are observed and their values are such that $D = d$. Since a large value of D would appear to be inconsistent with the null hypothesis that F is the underlying distribution, it follows that the p-value for this data set is given by

$$p\text{-value} = P_F\{D \geq d\}$$

where we have written P_F to make explicit that this probability is to be computed under the assumption that H_0 is correct (and so F is the underlying distribution).

The above p-value can be approximated by a simulation that is made easier by the following proposition, which shows that $P_F\{D \geq d\}$ does not depend on the underlying distribution F. This result enables us to estimate the p-value by doing the simulation with any continuous distribution F we choose [thus allowing us to use the uniform (0, 1) distribution].

Proposition:

$P_F\{D \geq d\}$ is the same for any continuous distribution F

Proof.

$$\begin{aligned} P_F\{D \geq d\} &= P_F\left\{\underset{x}{\text{Maximum}} \left|\frac{\#i\colon Y_i \leq x}{n} - F(x)\right| \geq d\right\} \\ &= P_F\left\{\underset{x}{\text{Maximum}} \left|\frac{\#i\colon F(Y_i) \leq F(x)}{n} - F(x)\right| \geq d\right\} \end{aligned}$$

$$= P\left\{\underset{x}{\text{Maximum}} \left|\frac{\#i\colon U_i \le F(x)}{n} - F(x)\right| \ge d\right\}$$

where $U_1, \ldots, U_n$ are independent uniform (0, 1) random variables: the first equality following because F is an increasing function and so $Y \le x$ is equivalent to $F(Y) \le F(x)$; and the second because of the result (whose proof is left as an exercise) that if Y has the continuous distribution F then the random variable $F(Y)$ is uniform on (0, 1).

Continuing the above, we see by letting $y = F(x)$ and noting that as x ranges from $-\infty$ to $+\infty$, $F(x)$ ranges from 0 to 1, that

$$P_F\{D \ge d\} = P\left\{\underset{0 \le y \le 1}{\text{Maximum}} \left|\frac{\#i\colon U_i \le y}{n} - y\right| \ge d\right\}$$

which shows that the distribution of D, when H_0 is true, does not depend on the actual distribution F. □

It follows from the above proposition that after the value of D is determined from the data, say, $D = d$, the p-value can be obtained by doing a simulation with the uniform (0, 1) distribution. That is, we generate a set of n random numbers $U_1, \ldots, U_n$ and then check whether or not the inequality

$$\underset{0 \le y \le 1}{\text{Maximum}} \left|\frac{\#i\colon U_i \le y}{n} - y\right| \ge d \tag{9.5}$$

is valid. This is then repeated many times and the proportion of times that it is valid is our estimate of the p-value of the data set. As noted earlier, the left side of the inequality (9.5) can be computed by ordering the random numbers and then using the identity

$$\text{Max} \left|\frac{\#i\colon U_i \le y}{n} - y\right|$$
$$= \text{Max}\left\{\frac{j}{n} - U_{(j)}, U_{(j)} - \frac{(j-1)}{n}, \qquad j = 1, \ldots, n\right\}$$

where $U_{(j)}$ is the jth smallest value of $U_1, \ldots, U_n$. For example, if $n = 3$ and $U_1 = .7$, $U_2 = .6$, $U_3 = .4$, then $U_{(1)} = .4$, $U_{(2)} = .6$, $U_{(3)} = .7$ and the value of D for this data set is

$$D = \text{Max}\{\tfrac{1}{3} - .4, \tfrac{2}{3} - .6, 1 - .7, .4, .6 - \tfrac{1}{3}, .7 - \tfrac{2}{3}\} = .4$$

Program 9-3 uses simulation to approximate the p-value for the Kolmogorov–Smirnov test quantity D. The user must input n, the size of the data set; d, the value of D based on the data; and the required number of simulation runs. The program generates the ordered values of a set of random numbers by using the approach outlined in Problem 8.

Example 9c Suppose we want to test the hypothesis that a given population distribution is exponential with mean 100; that is, $F(x) = 1 - e^{-x/100}$. If the (ordered) values from a sample of size 10 from this distribution are

$$66,\ 72,\ 81,\ 94,\ 112,\ 116,\ 124,\ 140,\ 145,\ 155$$

what conclusion can be drawn?

To answer the above, we first employ Equation (9.4) to compute the value of the Kolmogorov–Smirnov test quantity D. After some computation this gives the result $D = .4831487$. To obtain the approximate p-value we now run Program 9-3, which gives the following output:

```
RUN
THIS PROGRAM USES SIMULATION TO APPROXIMATE THE p-value OF THE
   KOLMOGOROV-SMIRNOV TEST
Random number seed (-32768 to 32767)? 4567
ENTER THE VALUE OF THE TEST QUANTITY
? .4831487
ENTER THE SAMPLE SIZE
? 10
ENTER THE DESIRED NUMBER OF SIMULATION RUNS
? 500
THE APPROXIMATE p-value IS .012
OK
```

Because the p-value is so low (it is extremely unlikely that the smallest of a set of 10 values from the exponential distribution with mean 100 would be as large as 66), the hypothesis would be rejected. ■

9.2. Goodness of Fit Tests When Some Parameters Are Unspecified

The Discrete Data Case

We can also perform a goodness of fit test of a null hypothesis that does not completely specify the probabilities $\{p_i, i = 1, \ldots, k\}$. For example, suppose we are interested in testing whether the daily number of traffic accidents in a certain region has a Poisson distribution with some unspecified mean. To test this hypothesis suppose that data is obtained over n days and let Y_i represent the number of accidents in day i, for $i = 1, \ldots, n$. To determine whether this data is consistent with the assumption of an underlying Poisson distribution we must first address the difficulty that, if the Poisson assumption is correct, these data can assume an infinite number of possible values. However, this is accomplished by breaking up the set of possible values into a finite number of, say, k regions and then seeing in which of the regions the n data points lie. For instance, if the geographical area of interest is small, and so there are not too many accidents in a day, we might say that the number of accidents in a given day falls in region i, $i = 1, 2, 3, 4, 5,$

when there are $i - 1$ accidents on that day, and in region 6 when there are 5 or more accidents. Hence, if the underlying distribution is indeed Poisson with mean λ, then

$$p_i = P\{Y = i - 1\} = \frac{e^{-\lambda}\lambda^{i-1}}{(i-1)!}, \qquad i = 1, 2, 3, 4, 5 \tag{9.6}$$

$$p_6 = 1 - \sum_{j=0}^{4} \frac{e^{-\lambda}\lambda^j}{j!}$$

Another difficulty we face in obtaining a goodness of fit test of the hypothesis that the underlying distribution is Poisson is that the mean value λ is not specified. Now, the intuitive thing to do when λ is unspecified is clearly to estimate its value from the data—call $\hat{\lambda}$ the estimate—and then compute the value of the test statistic

$$T = \sum_{i=1}^{k} \frac{(N_i - n\hat{p}_i)^2}{n\hat{p}_i}$$

where N_i is the number of the Y_j that fall in region i, and where $\hat{p}_i$ is the estimated probability, under H_0, that Y_j falls in region i, $i = 1, \ldots, k$, which is obtained by substituting $\hat{\lambda}$ for λ in the expression (9.6).

The above approach can be used whenever there are unspecified parameters in the null hypothesis that are needed to compute the quantities p_i, $i = 1, \ldots, k$. Suppose now that there are m such unspecified parameters. It can be proved that, for reasonable estimators of these parameters, when n is large the test quantity T has, when H_0 is true, approximately a chi-square distribution with k-1-m degrees of freedom. (In other words, one degree of freedom is lost for each parameter that needs to be estimated.)

If the test quantity takes on the value, say, $T = t$, then, using the above, the p-value can be approximated by

$$p\text{-value} \approx P\{X^2_{k\text{-}1\text{-}m} \geq t\}$$

where $X^2_{k\text{-}1\text{-}m}$ is a chi-square random variable with k-1-m degrees of freedom.

Example 9d Suppose that over a 30-day period there are 6 days in which no accidents occurred, 2 in which 1 accident occurred, 1 in which 2 accidents occurred, 9 in which 3 occurred, 7 in which 4 occurred, 4 in which 5 occurred, and 1 in which 8 occurred. To test whether this data is consistent with the hypothesis of an underlying Poisson distribution, note first that since there were a total of 87 accidents, the estimate of the mean of the Poisson distribution is

$$\hat{\lambda} = \frac{87}{30} = 2.9$$

Since the estimate of $P\{Y = i\}$ is thus $e^{-2.9}(2.9)^i/i!$, we obtain that with the six regions as given at the beginning of this section

$$\hat{p}_1 = .0500,\ \hat{p}_2 = .1596,\ \hat{p}_3 = .2312,\ \hat{p}_4 = .2237,$$
$$\hat{p}_5 = .1622,\ \hat{p}_6 = .1682$$

Using the data values $N_1 = 6$, $N_2 = 2$, $N_3 = 1$, $N_4 = 9$, $N_5 = 7$, $N_6 = 5$ we see that the value of the test statistic is

$$T = \sum_{i=1}^{6} \frac{(N_i - 30\hat{p}_i)^2}{30\hat{p}_i} = 19.887$$

To determine the p-value we run Program 9-1, which yields

$$p\text{-value} \approx P\{X_4^2 > 19.887\} = .0005$$

and so the hypothesis of an underlying Poisson distribution is rejected. ■

We can also use simulation to estimate the p-value. However, since the null hypothesis no longer completely specifies the probability model, the use of simulation to determine the p-value of the test statistic is somewhat trickier than before. The way it should be done is as follows.

(a) *The Model.* Suppose that the null hypothesis is that the data values $Y_1, \ldots, Y_n$ constitute a random sample from a distribution that is specified up to a set of unknown parameters $\theta_1, \ldots, \theta_m$. Suppose also that when this hypothesis is true, the possible values of the Y_i are $1, \ldots, k$.

(b) *The Initial Step.* Use the data to estimate the unknown parameters. Specifically, let $\hat{\theta}_j$ denote the value of the estimator of $\theta_j, j = 1, \ldots, m$. Now compute the value of the test statistic

$$T = \sum_{i=1}^{k} \frac{(N_i - n\hat{p}_i)^2}{n\hat{p}_i}$$

where N_i is the number of the data values that are equal to i, $i = 1, \ldots, k$, and $\hat{p}_i$ is the estimate of p_i that results when $\hat{\theta}_j$ is substituted for θ_j, for $j = 1, \ldots, m$. Let t denote the value of the test quantity T.

(c) *The Simulation Step.* We now do a series of simulations to estimate the p-value of the data. First note that all simulations are to be obtained by using the population distribution that results when the null hypothesis is true and θ_j is equal to its estimate $\hat{\theta}_j$, $j = 1, \ldots, m$, determined in step (b).

Simulate a sample of size n from the aforementioned population distribution and let $\hat{\theta}_j(\text{sim})$ denote the estimate of θ_j, $j = 1, \ldots, m$, based on the simulated data. Now determine the value of

$$T_{\text{sim}} = \sum_{i=1}^{k} \frac{[N_i - n\hat{p}_i(\text{sim})]^2}{n\hat{p}_i(\text{sim})}$$

where N_i is the number of the simulated data values equal to i, $i = 1, \ldots, k$, and $\hat{p}_i(\text{sim})$ is the value of p_i when θ_j is equal to $\theta_j(\text{sim})$, $j = 1, \ldots, m$.

The simulation step should then be repeated many times. The estimate of the p-value is then equal to the proportion of the values of T_{sim} that are at least as large as t.

Example 9e Let us reconsider Example 9d. The data presented in this example resulted in the estimate $\hat{\lambda} = 2.9$ and the test quantity value $T = 19.887$. The simulation step now consists of generating 30 independent Poisson random variables each having mean 2.9 and then computing the value of

$$T^* \equiv \sum_{i=1}^{6} \frac{(X_i - 30p_i^*)^2}{30p_i^*}$$

where X_i is the number of the 30 values that fall into region i, and p_i^* is the probability that a Poisson random variable with a mean equal to the average of the 30 generated values would fall into region i. This simulation step should be repeated many times and the estimated p-value is the proportion of times it results in a T^* at least as large as 19.887. ■

The Continuous Data Case

Now consider the situation where we want to test the hypothesis that the random variables $Y_1, \ldots, Y_n$ have the continuous distribution function $F_{\boldsymbol{\theta}}$, where $\boldsymbol{\theta} = (\theta_1, \ldots, \theta_m)$ is a vector of unknown parameters. For example, we might be interested in testing that the Y_j come from a normally distributed population. To employ the Kolmogorov–Smirnov test we first use the data to estimate the parameter vector $\boldsymbol{\theta}$, say, by the vector of estimators $\hat{\boldsymbol{\theta}}$. The value of the test statistic D is now computed by

$$D = \underset{x}{\text{Maximum}} \left|F_e(x) - F_{\hat{\boldsymbol{\theta}}}(x)\right|$$

where $F_{\hat{\boldsymbol{\theta}}}$ is the distribution function obtained from $F_{\boldsymbol{\theta}}$ when $\boldsymbol{\theta}$ is estimated by $\hat{\boldsymbol{\theta}}$.

If the value of the test quantity is $D = d$, then the p-value can be *roughly* approximated by $P_{F_{\hat{\boldsymbol{\theta}}}}\{D \geq d\} = P_U\{D \geq d\}$. That is, after determining the value of D, a rough approximation, which actually overestimates the p-value, is obtained. If this does not result in a small estimate for the p-value, then, as the hypothesis is not going to be rejected, we might as well stop. However, if this estimated p-value is small, then a more accurate way of using simulation to estimate the true p-value is necessary. We now describe how this should be done.

STEP 1: Use the data to estimate $\boldsymbol{\theta}$, say, by $\hat{\boldsymbol{\theta}}$. Compute the value of D as described above.

STEP 2: All simulations are to be done using the distribution $F_{\hat{\boldsymbol{\theta}}}$. Generate a sample of size n from this distribution and let $\hat{\boldsymbol{\theta}}(\text{sim})$ be the estimate of $\boldsymbol{\theta}$ based on this simulation run. Compute the value of

$$\underset{x}{\text{Maximum}}\ |F_{e,\text{sim}}(x) - F_{\hat{\boldsymbol{\theta}}(\text{sim})}(x)|$$

where $F_{e,\text{sim}}$ is the empirical distribution function of the simulated data; and note whether it is at least as large as d. Repeat this many times and use the proportion of times that this test quantity is at least as large as d as the estimate of the p-value.

9.3. The Two-Sample Problem

Suppose we have formulated a mathematical model for a service system which clears all its customers at the end of a day; and suppose that our model assumes that each day is probabilistically alike in that the probability laws for each day are identical and independent. Some of the individual assumptions of the model—such as, for example, that the service times are all independent with the common distribution G, or that the arrivals of customers constitute a Poisson process—can be individually tested by using the results of Sections 9.1 and 9.2. Suppose that none of these individual tests results in a particularly small p-value and so all the parts of the model, taken individually, do not appear to be inconsistent with the real data we have about the system. [We must be careful here in what we mean by a small p-value because, even if the model is correct, if we perform a large number of tests then, by chance, some of the resulting p-values may be small. For example, if we perform r separate tests on independent data, then the probability that at least one of the resulting p-values is as small as α is $1 - (1 - \alpha)^r$, which even for small α will become large as r increases.]

At this stage, however, we are still not justified in asserting that our model is correct and has been validated by the real data; for the totality of the model, including not only all the individual parts but also our assumptions about the ways in which these parts interact, may still be inaccurate. One way of testing the model in its entirety is to consider some random quantity that is a complicated function of the entire model. For example, we could consider the total amount of waiting time of all customers that enter the system in a given day. Suppose that we have observed the real system for

m days and let Y_i, $i = 1, \ldots, m$, denote the sum of these waiting times for day i. If we now simulate the proposed mathematical model for n days, we can let X_i, $i = 1, \ldots, n$, be the sum of the waiting times of all customers arriving on the (simulated) day i. Since the mathematical model supposes that all days are probabilistically alike and independent, it follows that all the random variables $X_1, \ldots, X_m$ have some common distribution, which we denote by F. Now if the mathematical model is an accurate representation of the real system, then the real data $Y_1, \ldots, Y_m$ also have the distribution F. That is, if the mathematical model is accurate, one should not be able to tell the simulated data apart from the real data. From this it follows that one way of testing the accuracy of the model in its entirety is to test the null hypothesis H_0 that $X_1, \ldots, X_n, Y_1, \ldots, Y_m$ are independent random variables having a common distribution. We now show how such a hypothesis can be tested.

Suppose we have two sets of data—$X_1, \ldots, X_n$ and $Y_1, \ldots, Y_m$—and we want to test the hypothesis H_0 that these $n + m$ random variables are all independent and identically distributed. This statistical hypothesis testing problem is called the two-sample problem.

To test H_0, order the $n + m$ values $X_1, \ldots, X_n, Y_1, \ldots, Y_m$ and suppose for the time being that all $n + m$ values are distinct and so the ordering is unique. Now for $i = 1, \ldots, n$, let R_i denote the rank of X_i among the $n + m$ data values; that is, $R_i = j$ if X_i is the jth smallest among the $n + m$ values. The quantity

$$R = \sum_{i=1}^{n} R_i$$

equal to the sum of the ranks of the first data set, is used as our test quantity. (Either of the two data sets can be considered as the "first" set.)

If R is either very large (indicating that the first data set tends to be larger than the second) or very small (indicating the reverse), then this would be strong evidence against the null hypothesis. Specifically, if $R = r$, we reject the null hypothesis if either

$$P_{H_0}\{R \le r\} \quad \text{or} \quad P_{H_0}\{R \ge r\}$$

is very low. Indeed, the p-value of the test data which results in $R = r$ is given by

$$p\text{-value} = 2 \text{ Minimum } (P_{H_0}\{R \le r\}, P_{H_0}\{R \ge r\}) \tag{9.7}$$

[It is twice the minimum of the probabilities because we reject either if R is too small or too large. For example, suppose r_* and r^* were such that the probability, under H_0, of obtaining a value less (greater) than or equal to r_* (r^*) is .05. Since the probability of either event occurring is, under H_0, .1 it follows that if the outcome is r_* (or r^*) the p-value is .1.]

The hypothesis test resulting from the above p-value—that is, the test that calls for rejection of the null hypothesis when the p-value is sufficiently small—is called the *two-sample rank sum test*. (Other names that have also been used to designate this test are the Wilcoxon two-sample test and the Mann–Whitney two-sample test.)

Example 9f Suppose that direct observation of a system over 5 days has yielded that a certain quantity has taken on the successive values

$$342, 448, 504, 361, 453$$

whereas a 10-day simulation of a mathematical model proposed for the system has resulted in the following values:

$$186, 220, 225, 456, 276, 199, 371, 426, 242, 311$$

Because the five data values from the first set have ranks 8, 12, 15, 9, 13, it follows that the value of the test quantity is $R = 57$. ■

We can explicitly compute the p-value given in Equation (9.7) when n and m are not too large and all the data are distinct. To do so let

$$P_{n,m}(r) = P_{H_0}\{R \le r\}$$

Hence $P_{n,m}(r)$ is the probability that from two identically distributed data sets of sizes n and m, the sum of the ranks of the data values from the first set is less than or equal to r. We can obtain a recursive equation for these probabilities by conditioning on whether the largest data value comes from the first or the second set. If the largest value is indeed contained in the first data set, the sum of the ranks of this set equals $n + m$ (the rank of the largest value) plus the sum of the ranks of the other $n - 1$ values from this set when considered along with the m values from the other set. Hence, when the largest is contained in the first data set, the sum of the ranks of that set is less than or equal to r if the sum of the ranks of the remaining $n - 1$ elements is less than or equal to $r - n - m$, and this is true with probability $P_{n-1,m}(r - n - m)$. By a similar argument we can show that if the largest value is contained in the second set, the sum of the ranks of the first set is less than or equal to r with probability $P_{n,m-1}(r)$. Finally, since the largest value is equally likely to be any of the $n + m$ values, it follows that it is a member of the first set with probability $n/(n + m)$. Putting this together yields the following recursive equation:

$$P_{n,m}(r) = \frac{n}{n+m} P_{n-1,m}(r - n - m) + \frac{m}{n+m} P_{n,m-1}(r) \tag{9.8}$$

Starting with the boundary conditions

$$P_{1,0}(k) = \begin{cases} 0, & k \le 0 \\ 1, & k > 0 \end{cases} \quad \text{and} \quad P_{0,1}(k) = \begin{cases} 0, & k < 0 \\ 1, & k \ge 0 \end{cases}$$

Equation (9.8) can be recursively solved to obtain $P_{n,m}(r) = P_{H_0}\{R \leq r\}$ and $P_{n,m}(r - 1) = 1 - P_{H_0}\{R \geq r\}$.

Program 9-4 uses the recursion presented in Equation (9.8) to compute the p-value for the rank sum test. The inputs needed are the sizes of the first and second data sets and the sum of the ranks of the elements of the first set. Whereas either set can be designated as the first set, the program runs fastest if the first set is the one whose sum of ranks is smallest.

Example 9g Five days of observation of a system yielded the following values of a certain quantity of interest:

$$132,\ 104,\ 162,\ 171,\ 129$$

A 10-day simulation of a proposed model of this system yielded the values

$$107,\ 94,\ 136,\ 99,\ 114,\ 122,\ 108,\ 130,\ 106,\ 88$$

Suppose the formulated model implies that these daily values should be independent and have a common distribution. To determine the p-value that results from the above data, note first that R, the sum of the ranks of the first sample, is

$$R = 12 + 4 + 14 + 15 + 10 = 55$$

To determine the p-value we now run Program 9-4, which yields

```
RUN
THIS PROGRAM COMPUTES THE p-value FOR THE TWO-SAMPLE RANK SUM
   TEST
THIS PROGRAM WILL RUN FASTEST IF YOU DESIGNATE AS THE FIRST SAM-
   PLE THE SAMPLE HAVING THE SMALLER SUM OF RANKS
ENTER THE SIZE OF THE FIRST SAMPLE
? 5
ENTER THE SIZE OF THE SECOND SAMPLE
? 10
ENTER THE SUM OF THE RANKS OF THE FIRST SAMPLE
? 55
THE p-value IS .0752579
OK
```

■

The difficulty with employing the recursion (9.8) to compute the p-value is that the amount of computation needed grows enormously as the sample sizes increase. For example, if $n = m = 20$, even if we choose the test quantity to be the smaller sum of ranks—since the sum of all the ranks is $1 + 2 + \cdots + 40 = 820$, it is possible that the test statistic could have a value as large as 410. Hence there can be as many as $20 \times 20 \times 410 = 164{,}000$ values of $P_{n,m}(r)$ that would have to be computed to determine the p-value. Thus, for large samples, the use of the recursion provided by (9.8) may not be viable. Two different approximation methods that can be used in such cases are (a) a classical approach based on approximating the distribution of R and (b) simulation.

To use the classical approach for approximating the p-value we make use of the fact that under H_0 all possible orderings of the $n + m$ values are equally likely. Using this fact it is easy to show that

$$E_{H_0}[R] = n\frac{(n + m + 1)}{2}$$

$$\text{Var}_{H_0}(R) = nm\frac{(n + m + 1)}{12}$$

Now it can be shown that, under H_0, when n and m are large, R is approximately normally distributed. Hence, when H_0 is true,

$$\frac{R - n(n + m + 1)/2}{\sqrt{nm(n + m + 1)/12}} \quad \text{is approximately a unit normal.}$$

Since, for a normal random variable W, the minimum of $P\{W \leq r\}$ and $P\{W \geq r\}$ is the former when $r \leq E[W]$, and the latter otherwise, it thus follows that when n and m are not too small (both being greater than 7 should suffice), we can approximate the p-value of the test result $R = r$ by

$$p\text{-value} \approx \begin{cases} 2\,P\{Z < r^*\} & \text{if } r \leq n\dfrac{(n + m + 1)}{2} \\ 2\,\text{P}\{\text{Z} > \text{r}^*\} & \text{otherwise} \end{cases} \tag{9.9}$$

where

$$r^* = \frac{r - \dfrac{n(n + m + 1)}{2}}{\sqrt{\dfrac{nm(n + m + 1)}{12}}}$$

and where Z is a unit normal random variable.

Example 9h Let us see how well the classical approximation works for the data of Example 9g. In this case, since $n = 5$ and $m = 10$, we have that

$$\begin{aligned} p\text{-value} &= 2\,P_{H_0}\{R \geq 55\} \\ &\approx 2\,P\left\{Z \geq \frac{(55 - 40)}{\sqrt{\dfrac{50 \times 16}{12}}}\right\} \\ &= 2\,P\{Z \geq 1.8371\} \\ &= .066 \end{aligned}$$

which should be compared with the exact answer .075. ■

The p-value of the two-sample rank test can also be approximated by simulation. To see how this is accomplished, recall that if the observed value of the test quantity R is $R = r$, then the p-value is given by

$$p\text{-value} = 2 \text{ Minimum } (P_{H_0}\{R \geq r\}, P_{H_0}\{R \leq r\})$$

Now, under H_0, provided that all the $n + m$ data values are distinct, it follows that all orderings among these data values are equally likely, and thus the ranks of the first data set of size n have the same distribution as a random selection of n of the values $1, 2, \ldots, n + m$. Thus, under H_0, the probability distribution of R can be approximated by continually simulating a random subset of n of the integers $1, 2, \ldots, n + m$ and determining the sum of the elements in the subset. The value of $P_{H_0}\{R \leq r\}$ can be approximated by the proportion of simulations that result in a sum less than or equal to r; and the value of $P_{H_0}\{R \geq r\}$ by the proportion of simulations that result in a sum greater than or equal to r.

Program 9-5 approximates the p-value by performing the preceding simulation. It operates by continually summing the last n terms of a random permutation of the numbers $1, 2, \ldots, n + m$. (The method used to simulate a random permutation is explained in Example 3a of Chapter 3). The program runs most efficiently when the data set of smallest size is designated as the first sample.

Example 9i Using Program 9-5 with 1000 runs, for the data in Example 9g yields the following:

```
RUN
THIS PROGRAM APPROXIMATES THE p-value IN THE TWO-SAMPLE RANK SUM
   TEST BY A SIMULATION STUDY
Random number seed (-32768 to 32767)? 4566
ENTER THE SIZE OF THE FIRST SAMPLE
? 5
ENTER THE SIZE OF THE SECOND SAMPLE
? 10
ENTER THE SUM OF THE RANKS OF THE FIRST SAMPLE
? 55
ENTER THE DESIRED NUMBER OF SIMULATION RUNS
? 1000
THE APPROXIMATE p-value IS .07
OK
```

■

The above analysis supposes that all the $n + m$ data values are distinct. When certain of the values have a common value, one should take as the rank of a data value the average of the ranks of the values equal to it. For example, if the first data set is 2, 3, 4 and the second 3, 5, 7, then the sum of the ranks of the first set is $1 + 2.5 + 4 = 7.5$. The p-value should be approximated by using the normal approximation via Equation (9.9).

A generalization of the two-sample problem is the multisample problem, where one has the following m data sets,

$$\begin{array}{cccc} X_{1,1}, & X_{1,2}, & \ldots, & X_{1,n_1} \\ X_{2,1}, & X_{2,2}, & \ldots, & X_{2,n_2} \\ \vdots & \vdots & \vdots & \vdots \\ X_{m,1}, & X_{m,2}, & \ldots, & X_{m,n_m} \end{array}$$

and we are interested in testing the null hypothesis H_0 that all the $n = \sum_{i=1}^{m} n_i$ random variables are independent and have a common distribution. A generalization of the two-sample rank test, called the multisample rank test (or often referred to as the Kruskal–Wallis test), is obtained by first ranking all the n data values. Then let R_i, $i = 1, \ldots, m$, denote the sum of the ranks of all the n_i data values from the ith set. (Note that with this notation R_i is a sum of ranks and not an individual rank as previously.) Since, under H_0, all orderings are equally likely (provided all the data values are distinct), it follows exactly as before that

$$E[R_i] = n_i \frac{(n+1)}{2}$$

Using the above, the multisample rank sum test is based on the test quantity

$$R = \frac{12}{n(n+1)} \sum_{i=1}^{m} \frac{[R_i - n_i(n+1)/2]^2}{n_i}$$

Since small values of R indicate a good fit to H_0, the test based on the quantity R rejects H_0 for sufficiently large values of R. Indeed, if the observed value of R is $R = y$, the p-value of this result is given by

$$p\text{-value} = P_{H_0}\{R \geq y\}$$

This value can be approximated by using the result that for large values of $n_1, \ldots, n_m$, R has approximately a chi-square distribution with $m - 1$ degrees of freedom [this latter result being the reason why we include the term $12/n(n + 1)$ in the definition of R]. Hence, if $R = y$,

$$p\text{-value} \approx P\{X^2_{m-1} \geq y\}$$

Simulation can also be used to evaluate the p-value (see Problem 14).

Even when the data values are not all distinct, the above approximation for the p-value should be used. In computing the value of R the rank of an individual data value should be, as before, the average of all the ranks of the data equal to it.

9.4. Validating the Assumption of a Nonhomogeneous Poisson Process

Consider a mathematical model which supposes that the daily arrivals to a system occur in accordance with a nonhomogeneous Poisson process, with the arrival process from day to day being independent and having a common, but unspecified, intensity function.

To validate such an assumption, suppose that we observe the system over r days, noting the arrival times. Let N_i, $i = 1, \ldots, r$, denote the number of arrivals on day i; and note that if the arrival process is indeed a nonhomogeneous Poisson process, then these quantities are independent Poisson random variables with the same mean. Now whereas this consequence could be tested by using the goodness of fit approach, as is done in Example 9a, we present an alternative approach that is sometimes more efficient. This alternative approach is based on the fact that the mean and variance of a Poisson random variable are equal. Hence, if the N_i are indeed a sample from a Poisson distribution, the sample mean

$$\overline{N} = \sum_{i=1}^{r} \frac{N_i}{r}$$

and the sample variance

$$S^2 = \sum_{i=1}^{r} \frac{(N_i - \overline{N})^2}{(r - 1)}$$

should be roughly equal. Motivated by this, we base our test of the hypothesis $H_0: N_i$ are independent Poisson random variables with a common mean on the test quantity

$$T = \frac{S^2}{\overline{N}} \tag{9.10}$$

Because either a very small or very large value of T would be inconsistent with H_0, the p-value for the outcome $T = t$ would be

$$p\text{-value} = 2 \text{ Minimum } (P_{H_0}\{T \leq t\}, P_{H_0}\{T \geq t\})$$

However, since H_0 does not specify the mean of the Poisson distribution, we cannot immediately compute the above probabilities; rather, we must first use the observed data to estimate the mean. By using the estimator $\overline{N}$, it follows that if the observed value of $\overline{N}$ is $\overline{N} = m$, the p-value can be approximated by

$$p\text{-value} \approx 2 \text{ Minimum } (P_m\{T \leq t\}, P_m\{T \geq t\})$$

where T is defined by Equation (9.10) with $N_1, \ldots, N_r$ being independent Poisson random variables each with mean m. We can now approximate $P_m\{T \le t\}$ and $P_m\{T \ge t\}$ via a simulation. That is, we continually generate r independent Poisson random variables with mean m and compute the resulting value of T. The proportion of these for which $T \le t$ is our estimate of $P\{T \le t\}$, and the proportion for which $T \ge t$ is our estimate of $P\{T \ge t\}$.

If the above p-value is quite small, we reject the null hypothesis that the daily arrivals constitute a nonhomogeneous Poisson process. However, if the p-value is not small, this only implies that the assumption that the number of arrivals each day has a Poisson distribution is a viable assumption and does not by itself validate the stronger assumption that the actual arrival pattern (as determined by the nonhomogeneous intensity function) is the same from day to day. To complete our validation we must now consider the actual arrival times for each of the r days observed. Suppose that the arrival times on day j, $j = 1, \ldots, r$, are $X_{j,1}, X_{j,2}, \ldots, X_{j,N_j}$. Now if the arrival process is indeed a nonhomogeneous Poisson process, it can be shown that each of these r sets of arrival times constitutes a sample from a common distribution. That is, under the null hypothesis, the r sets of data $X_{j,1}, \ldots, X_{j,N_j}$, $j = 1, \ldots, r$, are all independent random variables from a common distribution.

The above consequence, however, can be tested by the multisample rank test given in Section 9.3. That is, first rank all the $N \equiv \sum_{j=1}^{r} N_j$ data values, and then let R_j denote the sum of the ranks of all the N_j data values from the ith set. The test quantity

$$R = \frac{12}{N(N+1)} \sum_{j=1}^{r} \frac{\left(R_j - N_j \dfrac{(N+1)}{2}\right)^2}{N_j}$$

can now be employed by using the fact that, when H_0 is true, R has approximately a chi-square distribution with $r - 1$ degrees of freedom. Hence, if the observed value of R is $R = y$, the resulting p-value can be approximated by

$$\begin{aligned} p\text{-value} &= 2 \text{ Minimum } (P_{H_0}\{R \le y\}, P_{H_0}\{R \ge y\}) \\ &\approx 2 \text{ Minimum } (P\{X_{r-1}^2 \le y\}, 1 - P\{X_{r-1}^2 \le y\}) \end{aligned}$$

where X_{r-1}^2 is a chi-square random variable with $r - 1$ degrees of freedom. (Of course, we could also approximate the p-value by a simulation.) If the above p-value, along with the previous p-value considered, is not too small, we may conclude that the data are not inconsistent with our assumption that daily arrivals constitute a nonhomogeneous Poisson process.

Example 9j Suppose that the daily times at which deliveries are made at a certain plant are noted over 5 days. During this time the number of deliveries during each of the days are as follows:

$$18, 24, 16, 19, 25$$

Suppose also that when the 102 delivery times are ranked according to the time of day they arrived, the sums of the ranks of the deliveries from each day are

$$1010, 960, 1180, 985, 1118$$

Using the above data, let us test the hypothesis that the daily arrival process of deliveries is a nonhomogeneous Poisson process.

We first test that the first data set of the daily number of deliveries consists of a set of five independent and identically distributed Poisson random variables. Now the sample mean and sample variance are equal to

$$\overline{N} = 20.4 \qquad \text{and} \qquad S^2 = 15.3$$

and so the value of the test quantity is $T = .75$. To determine the approximate p-value of the test that the N_i are independent Poisson random variables, we then simulated 500 sets of five Poisson random variables with mean 20.4 and then computed the resulting value of $T = S^2/\overline{N}$. The output of this simulation indicated a p-value of approximately .84, and so it is clear that the assumption that the number of daily deliveries are independent Poisson random variables having a common mean is consistent with the data.

To continue our test of the null hypothesis of a nonhomogeneous Poisson process, we compute the value of the test quantity R, which is seen to be equal to 14.425. Because the probability that a chi-square random variable with four degrees of freedom is as large as 14.425 is (from Program 9-1) .006, it follows that the p-value is .012. For such a small p-value we must reject the null hypothesis. ■

If we wanted to test the assumption that the daily arrival process constituted a *homogeneous* Poisson process, we would proceed as above and first test the hypothesis that the number of arrivals each day are independent and identically distributed Poisson random variables. If the hypothesis remains plausible after performing this test, we again continue as in the nonhomogeneous case by considering the actual set of $N = \sum_{j=1}^{r} N_j$ arrival times. However, we now use the result that, under a homogeneous Poisson process, it can be shown that, given the number of arrivals in a day, the actual arrivals are independently and uniformly distributed over $(0, T)$, where T is the length of a day. This consequence, however, can be tested by the Kolmogorov–Smirnov goodness of fit test presented in Section 9.1. That is, if the arrivals constitute a homogeneous Poisson process, the N random variables $X_{j,i}$, $i = 1, \ldots, N_j$, $j = 1, \ldots, r$, where $X_{j,i}$ represents the ith arrival time on day j, can be regarded as constituting a set of N independent and uniformly distributed random variables over $(0, T)$. Hence, if we define the empirical distribution function F_e by letting $F_e(x)$ be the proportion of the N data values that are less than or equal to x—that is,

$$F_e(x) = \sum_{j=1}^{r} \sum_{i=1}^{N_j} \frac{I_{j,i}}{N}$$

where

$$I_{j,i} = \begin{cases} 1 & \text{if } X_{j,i} \leq x \\ 0 & \text{otherwise} \end{cases}$$

then the value of the test quantity is

$$D = \underset{0 \leq x \leq T}{\text{Maximum}} \left| F_e(x) - \frac{x}{T} \right|$$

Once the value of the test statistic D is determined, we can then find the resulting p-value by simulation, as is shown in Section 9.1.

If the hypothesis of a nonhomogeneous Poisson process is shown to be consistent with the data, we face the problem of estimating the intensity function $\lambda(t)$, $0 \leq t \leq T$, of this process. [In the homogeneous case the obvious estimator is $\lambda(t) = \hat{\lambda}/T$, where $\hat{\lambda}$ is the estimate of the mean number of arrivals in a day of length T.] To estimate the intensity function, order the $N = \sum_{j=1}^{r} N_j$ daily arrival times. Let $y_0 = 0$, and for $k = 1, \ldots, N$, let y_k denote the kth smallest of these N arrival times. Because there has been a total of 1 arrival over r days within the time interval (y_{k-1}, y_k), $k = 1, \ldots, N$, a reasonable estimate of $\lambda(t)$ would be

$$\hat{\lambda}(t) = \frac{1}{r(y_k - y_{k-1})} \qquad \text{for } y_{k-1} < t < y_k$$

(To understand the above estimator, note that if $\hat{\lambda}(t)$ were the intensity function, the expected number of daily arrivals that occur at a time point t such that $y_{k-1} < t \leq y_k$ would be given by

$$E[N(y_k) - N(y_{k-1})] = \int_{y_{k-1}}^{y_k} \hat{\lambda}(t)\, dt = \frac{1}{r}$$

and hence the expected number of arrivals within that interval over r days would be 1, which coincides with the actual observed number of arrivals in that interval.)

Problems

1. According to the Mendelian theory of genetics, a certain garden pea plant should produce either white, pink, or red flowers, with respective probabilities $\frac{1}{4}$, $\frac{1}{2}$, $\frac{1}{4}$. To test this theory a sample of 564 peas were studied with the result that 141 produced white, 291 produced pink, and 132 produced red flowers. Approximate the p-value of this data set
(a) by using the chi-square approximation, and
(b) by using a simulation.

2. To ascertain whether a certain die was fair, 1000 rolls of the die were recorded, with the result that the number of times the die landed i, $i = 1, 2, 3, 4, 5, 6$ were, respectively, 158, 172, 164, 181, 160, 165. Approximate the p-value of the test that the die was fair
(a) by using the chi-square approximation, and
(b) by using simulation.

3. Approximate the p-value of the hypothesis that the following 10 values are random numbers: .12, .18, .06, .33, .72, .83, .36, .27, .77, .74.

4. Approximate the p-value of the hypothesis that the following data set of 14 points is a sample from a uniform distribution over (50, 200):

164, 142, 110, 153, 103, 52, 174, 88, 178, 184, 58, 62, 132, 128

5. Approximate the p-value of the hypothesis that the following 13 data values come from an exponential distribution with mean 50:

86, 133, 75, 22, 11, 144, 78, 122, 8, 146, 33, 41, 99

6. Approximate the p-value of the test that the following data come from a binomial distribution with parameters $(8, p)$, where p is unknown:

6, 7, 3, 4, 7, 3, 7, 2, 6, 3, 7, 8, 2, 1, 3, 5, 8, 7

7. Approximate the p-value of the test that the following data set comes from an exponentially distributed population: 122, 133, 106, 128, 135, 126.

8. To generate the ordered values of n random numbers we could generate n random numbers and then order, or sort, them. Another approach makes use of the result that given that the $(n + 1)$st event of a Poisson process occurs at time t, the first n event times are distributed as the set of ordered values of n uniform $(0, t)$ random variables. Using this result, explain why, in the following algorithm, $y_1, \ldots, y_n$ denote the ordered values of n random numbers.

Generate $n + 1$ random numbers $U_1, \ldots, U_{n+1}$

$X_i = -\log U_i, \qquad i = 1, \ldots, n + 1$

$$t = \sum_{i=1}^{n+1} X_i, \qquad c = \frac{1}{t}$$

$$y_i = y_{i-1} + cX_i, \qquad i = 1, \ldots, n \text{ (with } y_0 = 0)$$

9. Generate the values of 10 independent exponential random variables each having mean 1. Then, based on the Kolmogorov–Smirnov test quantity, approximate the p-value of the test that the data do indeed come from an exponential distribution with mean 1.

10. An experiment designed to compare two treatments against corrosion yielded the following data (representing the maximum depth of pits in units of one-thousandth of an inch) in pieces of wire subjected to one or the other of the two treatments:

Treatment 1:	65.2	67.1	69.4	78.4	74.0	80.3
Treatment 2:	59.4	72.1	68.0	66.2	58.5	

Compute the exact p-value of this data set when testing the hypothesis that the two treatments have identical results.

11. In Problem 10, compute the approximate p-value based on
(a) the normal approximation, and
(b) a simulation.

12. Fourteen cities, of roughly equal size, are chosen for a traffic safety study. Seven of them are randomly chosen, and in these cities a series of newspaper articles dealing with traffic safety are run over a 1-month period. The number of traffic accidents reported in the month following this campaign are as follows:

Treatment group:	19	31	39	45	47	66	75
Control group:	28	36	44	49	52	72	72

Determine the exact p-value when testing the hypothesis that the articles have not had any effect.

13. Approximate the p-value in Problem 12
(a) by using the normal approximation, and
(b) by using a simulation.

14. Explain how simulation can be employed to approximate the p-value in the multisample problem—that is, when testing that a set of m samples all come from the same probability distribution.

15. Consider the following data resulting from three samples:

Sample 1:	121	144	158	169	194	211	242
Sample 2:	99	128	165	193	242	265	302
Sample 3:	129	134	137	143	152	159	170

Compute the approximate p-value of the test that all the data come from a single probability distribution
(a) by using the chi-square approximation, and
(b) by using a simulation.

16. The number of daily arrivals over a 7-day interval are as follows:

122, 118, 120, 116, 125, 119, 124, 130

Do you think the daily arrivals could be independent and identically distributed as nonhomogeneous Poisson processes?

17. Over an interval of length 100 there have been 18 arrivals at the following times:

12, 20, 33, 44, 55, 56, 61, 63, 66, 70, 73, 75, 78, 80, 82, 85, 87, 90

Approximate the p-value of the test that the arrival process is a (homogeneous) Poisson process.

CHAPTER 10

Some Advanced Topics

Introduction

In this chapter we present some additional topics that are somewhat more specialized than those considered in earlier chapters. In Section 10.1 we present a highly efficient technique, called the alias method, for generating discrete random variables. In Section 10.2 we define and then present a method for simulating a two-dimensional Poisson process. In Section 10.3 we consider the Markov chain simulation method, which is quite useful when dealing with random "combinatorial" elements. Finally, Section 10.4 considers how to efficiently estimate probabilities and expectations concerning first passage times by making use of a powerful variable called the total hazard.

10.1. The Alias Method for Generating Discrete Random Variables†

In this section we study a technique for generating discrete random variables which, although requiring some set-up time, is very fast to implement.

In what follows, the quantities $\mathbf{P}$, $\mathbf{P}^{(k)}$, $\mathbf{Q}^{(k)}$, $k \leq n - 1$ represent probability mass functions on the integers $1, 2, \ldots, n$—that is, they are n-vectors of nonnegative numbers summing to 1. In addition, the vector $\mathbf{P}^{(k)}$ has at most k nonzero components, and each of the $\mathbf{Q}^{(k)}$ has at most two nonzero components. We show that any probability mass function $\mathbf{P}$ can be represented as an equally weighted mixture of $n - 1$ probability mass functions $\mathbf{Q}$ (each having at most two nonzero components). That is, we show, for suitably defined $\mathbf{Q}^{(1)}, \ldots, \mathbf{Q}^{(n-1)}$, that $\mathbf{P}$ can be expressed as

†This section is taken from S. Ross, *Introduction to Probability Models*, 4th ed., New York: Academic Press, 1989, with permission.

$$\mathbf{P} = \frac{1}{n-1} \sum_{k=1}^{n-1} \mathbf{Q}^{(k)} \tag{10.1}$$

As a prelude to presenting the method for obtaining this representation, we need the following simple lemma whose proof is left as an exercise.

Lemma 10.1:

Let $\mathbf{P} = \{P_i,\ i = 1, \ldots, n\}$ *denote a probability mass function, then*

(a) there exists an i, $1 \le i \le n$, such that $P_i < 1/(n-1)$, and
(b) for this i there exists a j, $j \ne i$, such that $P_i + P_j \ge 1/(n-1)$.

Before presenting the general technique for obtaining the representation (10.1), let us illustrate it by an example.

Example 10a Consider the three-point distribution $\mathbf{P}$ with $P_1 = \frac{7}{16}$, $P_2 = \frac{1}{2}$, $P_3 = \frac{1}{16}$. We start by choosing i and j satisfying the conditions of Lemma 10.1. Since $P_3 < \frac{1}{2}$ and $P_3 + P_2 > \frac{1}{2}$, we can work with $i = 3$ and $j = 2$. We now define a two-point mass function $\mathbf{Q}^{(1)}$, putting all its weight on 3 and 2 and such that $\mathbf{P}$ is expressible as an equally weighted mixture between $\mathbf{Q}^{(1)}$ and a second two-point mass function $\mathbf{Q}^{(2)}$. In addition, all the mass of point 3 is contained in $\mathbf{Q}^{(1)}$. As we have

$$P_j = \tfrac{1}{2}\,(Q_j^{(1)} + Q_j^{(2)}), \qquad j = 1, 2, 3 \tag{10.2}$$

and, by the above, $Q_3^{(2)}$ is supposed to equal 0, we must therefore take

$$Q_3^{(1)} = 2P_3 = \tfrac{1}{8}, \qquad Q_2^{(1)} = 1 - Q_3^{(1)} = \tfrac{7}{8}, \qquad Q_1^{(1)} = 0$$

To satisfy (10.2), we must then set

$$Q_3^{(2)} = 0, \qquad Q_2^{(2)} = 2P_2 - \tfrac{7}{8} = \tfrac{1}{8}, \qquad Q_1^{(2)} = 2P_1 = \tfrac{7}{8}$$

Hence we have the desired representation in this case. Suppose now that the original distribution was the following four-point mass function:

$$P_1 = \tfrac{7}{16}, \qquad P_2 = \tfrac{1}{4}, \qquad P_3 = \tfrac{1}{8}, \qquad P_4 = \tfrac{3}{16}$$

Now $P_3 < \frac{1}{3}$ and $P_3 + P_1 > \frac{1}{3}$. Hence our initial two-point mass function—$\mathbf{Q}^{(1)}$—concentrates on points 3 and 1 (giving no weight to 2 and 4). Because the final representation gives weight $\frac{1}{3}$ to $\mathbf{Q}^{(1)}$ and in addition the other $\mathbf{Q}^{(j)}$, $j = 2, 3$, do not give any mass to the value 3, we must have that

$$\tfrac{1}{3}\, Q_3^{(1)} = P_3 = \tfrac{1}{8}$$

Hence

$$Q_3^{(1)} = \tfrac{3}{8}, \qquad Q_1^{(1)} = 1 - \tfrac{3}{8} = \tfrac{5}{8}$$

Also, we can write

$$\mathbf{P} = \tfrac{1}{3}\, \mathbf{Q}^{(1)} + \tfrac{2}{3}\, \mathbf{P}^{(3)}$$

where $\mathbf{P}^{(3)}$, to satisfy the above, must be the vector

$$\begin{aligned} \mathbf{P}_1^{(3)} &= \tfrac{3}{2}\,(P_1 - \tfrac{1}{3}\, Q_1^{(1)}) = \tfrac{11}{32} \\ \mathbf{P}_2^{(3)} &= \tfrac{3}{2}\, P_2 = \tfrac{3}{8} \\ \mathbf{P}_3^{(3)} &= 0 \\ \mathbf{P}_4^{(3)} &= \tfrac{3}{2}\, P_4 = \tfrac{9}{32} \end{aligned}$$

Note that $\mathbf{P}^{(3)}$ gives no mass to the value 3. We can now express the mass function $\mathbf{P}^{(3)}$ as an equally weighted mixture of two-point mass functions $\mathbf{Q}^{(2)}$ and $\mathbf{Q}^{(3)}$, and we end up with

$$\begin{aligned} \mathbf{P} &= \tfrac{1}{3}\, \mathbf{Q}^{(1)} + \tfrac{2}{3}(\tfrac{1}{2}\, \mathbf{Q}^{(2)} + \tfrac{1}{2}\, \mathbf{Q}^{(3)}) \\ &= \tfrac{1}{3}(\mathbf{Q}^{(1)} + \mathbf{Q}^{(2)} + \mathbf{Q}^{(3)}) \end{aligned}$$

(We leave it as an exercise for the reader to fill in the details.) ■

The above example outlines the following general procedure for writing the n-point mass function $\mathbf{P}$ in the form (10.1), where each of the $\mathbf{Q}^{(i)}$ are mass functions giving all their mass to at most two-points. To start, we choose i and j satisfying the conditions of Lemma 10.1. We now define the mass function $\mathbf{Q}^{(1)}$ concentrating on the points i and j and which contain all the mass for point i by noting that in the representation (10.1) $Q_i^{(k)} = 0$ for $k = 2, \ldots, n-1$, implying that

$$Q_i^{(1)} = (n-1)P_i \qquad \text{and so} \qquad Q_j^{(1)} = 1 - (n-1)P_i$$

Writing

$$\mathbf{P} = \frac{1}{n-1}\mathbf{Q}^{(1)} + \frac{n-2}{n-1}\mathbf{P}^{(n-1)} \tag{10.3}$$

where $\mathbf{P}^{(n-1)}$ represents the remaining mass, we see that

$$P_i^{(n-1)} = 0$$

$$P_j^{(n-1)} = \frac{n-1}{n-2}\left(P_j - \frac{1}{n-1}Q_j^{(1)}\right) = \frac{n-1}{n-2}\left(P_i + P_j - \frac{1}{n-1}\right)$$

$$P_k^{(n-1)} = \frac{n-1}{n-2}P_k, \qquad k \neq i \quad \text{or} \quad j$$

That the above is indeed a probability mass function is easily checked—for example, the nonnegativity of $P_j^{(n-1)}$ follows from the fact that j was chosen so that $P_i + P_j \geq 1/(n-1)$.

We may now repeat the above procedure on the $(n-1)$ point probability mass function $\mathbf{P}^{(n-1)}$ to obtain

$$\mathbf{P}^{(n-1)} = \frac{1}{n-2}\mathbf{Q}^{(2)} + \frac{n-3}{n-2}\mathbf{P}^{(n-2)}$$

and thus from (10.3) we have

$$\mathbf{P} = \frac{1}{n-1}\mathbf{Q}^{(1)} + \frac{1}{n-1}\mathbf{Q}^{(2)} + \frac{n-3}{n-1}\mathbf{P}^{(n-2)}$$

We now repeat the procedure on $P^{(n-2)}$ and so on until we finally obtain

$$\mathbf{P} = \frac{1}{n-1}(\mathbf{Q}^{(1)} + \cdots + \mathbf{Q}^{(n-1)})$$

In this way we are able to represent $\mathbf{P}$ as an equally weighted mixture of $n-1$ two-point mass functions. We can now easily simulate from $\mathbf{P}$ by first generating a random integer N equally likely to be either $1, 2, \ldots, n-1$. If the resulting value N is such that $\mathbf{Q}^{(N)}$ puts positive weight only on the points i_N and j_N, we can set X equal to i_N if a second random number is less than $Q_{i_N}^{(N)}$ and equal to j_N otherwise. The random variable X will have probability mass function $\mathbf{P}$. That is, we have the following procedure for simulating from $\mathbf{P}$.

STEP 1: Generate U_1 and set $N = 1 + \text{Int}[(n-1)U_1]$

STEP 2: Generate U_2 and set

$$X = \begin{cases} i_N & \text{if } U_2 < Q_{i_N}^{(N)} \\ j_N & \text{otherwise} \end{cases}$$

REMARKS

1. The above is called the alias method because by a renumbering of the $\mathbf{Q}$'s we can always arrange things so that for each k, $Q_k^{(k)} > 0$. (That is, we can arrange things so that the kth two-point mass function gives positive weight to the value k.) Hence the procedure calls for simulating N, equally likely to be $1, 2, \ldots, n-1$, and then if $N = k$ it either accepts k as the value of X, or it accepts for the value of X the "alias" of k (namely, the other value that $\mathbf{Q}^{(k)}$ gives positive weight).

2. Actually, it is not necessary to generate a new random number in Step 2. Because $N-1$ is the integer part of $(n-1)U_1$, it follows that the remainder $(n-1)U_1 - (N-1)$ is independent of U_1 and is uniformly distributed on $(0, 1)$. Hence, rather than generating a new random number U_2 in Step 2, we can use $(n-1)U_1 - (N-1)$. ■

10.2. Simulating a Two-Dimensional Poisson Process

A process consisting of randomly occurring points in the plane is said to constitute a two-dimensional Poisson process having rate λ, $\lambda > 0$, if

1. the number of points occurring in any given region of area A is Poisson distributed with mean λA.
2. the number of points occurring in disjoint regions are independent.

For a given fixed point $\boldsymbol{O}$ in the plane, we now show how to simulate points, according to a two-dimensional Poisson process with rate λ, that occur in a circular region of radius r centered at $\boldsymbol{O}$.

Let $C(a)$ denote the circle of radius a centered at $\boldsymbol{O}$, and note that, from Condition 1, the number of points in $C(a)$ is Poisson distributed with mean πa^2. Let R_i, $i \geq 1$, denote the distance from the origin $\boldsymbol{O}$ to its ith nearest point (Figure 10.1). Then

$$\begin{aligned} P\{\pi R_1^2 > x\} &= P\{R_1 > \sqrt{x/\pi}\} \\ &= P\left\{\text{no points in } C\left(\sqrt{\frac{x}{\pi}}\right)\right\} \\ &= e^{-\lambda x} \end{aligned}$$

where the last equality uses the fact that the area of $C(\sqrt{x/\pi})$ is x. Also, with $C(b) - C(a)$ denoting the region between $C(b)$ and $C(a)$, $a < b$, we have

$$P\{\pi R_2^2 - \pi R_1^2 > x | R_1 = a\} = P\left\{R_2 > \sqrt{\frac{(x + \pi R_1^2)}{\pi}} \,\middle|\, R_1 = a\right\}$$

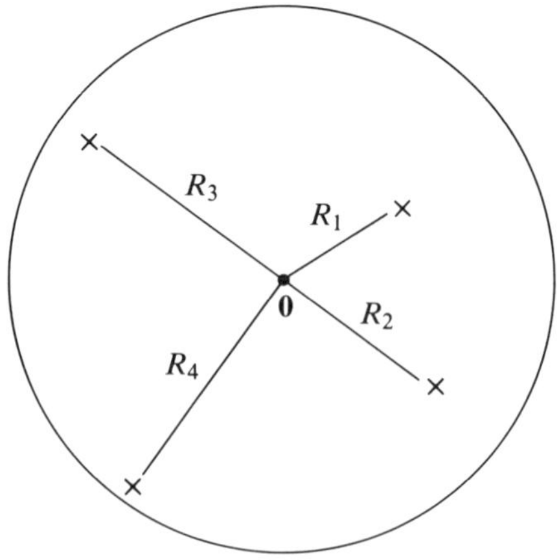

Figure 10.1

$$= P\left\{\text{no points in } C\left(\sqrt{\frac{(x+\pi a^2)}{\pi}}\right) - C(a) \middle| R_1 = a\right\}$$

$$= P\left\{\text{no points in } C\left(\sqrt{\frac{(x+\pi a^2)}{\pi}}\right) - C(a)\right\} \qquad \text{by Condition 2}$$

$$= e^{-\lambda x}$$

In fact, the same argument can be repeated continually to obtain the following proposition.

Proposition:

With $R_0 = 0$, $\pi R_i^2 - \pi R_{i-1}^2$, $i \geq 1$, are independent exponential random variables each having rate λ.

In other words, the amount of area that need be traversed to encounter a Poisson point is exponential with rate λ. Since, by symmetry, the respective angles of the Poisson points are independent and uniformly distributed over $(0, 2\pi)$, we thus have the following algorithm for simulating the Poisson process over a circular region of radius r about $\boldsymbol{O}$.

STEP 1: Generate independent exponentials with rate λ, $X_1, X_2, \ldots,$ stopping at

$$N = \text{Min}\{n: X_1 + \cdots + X_n > \pi r^2\}$$

STEP 2: If $N = 1$ stop; there are no points in $C(r)$. Otherwise, for $i = 1, \ldots, N-1$, set

$$R_i = \sqrt{\frac{X_1 + \cdots + X_i}{\pi}}$$

(that is, $\pi R_i^2 = X_1 + \cdots + X_i$).

STEP 3: Generate random numbers $U_1, \ldots, U_{N-1}$.

STEP 4: The polar coordinates of the $N-1$ Poisson points are

$$(R_i, 2\pi U_i), \qquad i = 1, \ldots, N-1$$

The above algorithm can be considered as the fanning out from a circle centered at **O** with a radius that expands continuously from 0 to r. The successive radii at which points are encountered is simulated by using the result that the additional area necessary to explore until one encounters another point is always exponentially distributed with rate λ. This fanning out technique can also be used to simulate the process over noncircular regions.

For example, consider a nonnegative function $f(x)$ and suppose that we are interested in simulating the Poisson process in the region between the x-axis and the function f (Figure 10.2) with x going from 0 to T. To do so, we can start at the left-hand edge and fan vertically to the right by considering the successive areas encountered. Specifically, if $X_1 < X_2 < \cdots$ denote the successive projections on the Poisson process on the x-axis, it follows in exactly the same manner as before that (with $X_0 = 0$)

$$\int_{X_{i-1}}^{X_i} f(x)\,dx, \qquad i = 1, \ldots, \text{ are independent exponentials with rate } \lambda$$

Hence we can simulate the above by generating independent exponential random variables with rate λ, $W_1, W_2, \ldots$ stopping at

$$N = \text{Min}\left\{n\colon\ W_1 + \cdots + W_n > \int_0^T f(x)\,dx\right\}$$

We should now determine $X_1, \ldots, X_{N-1}$ by using the equations

$$\int_0^{X_1} f(x)\,dx = W_1$$

$$\int_{X_1}^{X_2} f(x)\,dx = W_2$$

$$\vdots$$

$$\int_{X_{N-2}}^{X_{N-1}} f(x)\,dx = W_{N-1}$$

Because the projection on the y-axis of the point whose x-coordinate is X_i is clearly uniformly distributed over $(0, f(X_i))$, it thus follows that if we now

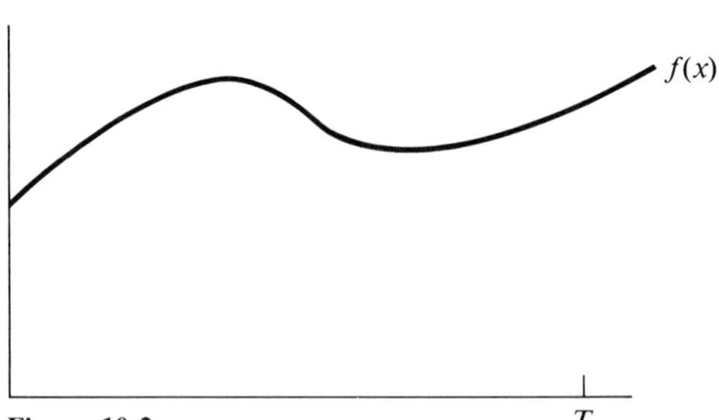

Figure 10.2

generate random numbers $U_1, \ldots, U_{N-1}$ then the simulated Poisson points are, in rectangular coordinates, $(X_i, U_i f(X_i))$, $i = 1, \ldots, N - 1$.

The above procedure is most useful when f is regular enough so that the above equations can be efficiently solved for the values of X_i. For example, if $f(x) = c$ (and so the region is a rectangle), we can express X_i as

$$X_i = \frac{W_1 + \cdots + W_i}{c}$$

and the Poisson points are

$$(X_i, cU_i), \qquad i = 1, \ldots, N - 1$$

10.3. The Markov Chain Simulation Method

Consider a collection of random variables $X_0, X_1, \ldots$. Interpret X_n as the "state of the system at time n," and suppose that the set of possible values of the X_n—that is, the possible states of the system—is the set $1, \ldots, N$. If there exists a set of numbers P_{ij}, $i, j = 1, \ldots, N$, such that whenever the process is in state i then, independent of the past states, the probability that the next state is j is P_{ij}, then we say that the collection $\{X_n, n \geq 0\}$ constitutes a *Markov chain* having transition probabilities P_{ij}, $i, j = 1, \ldots, N$. Since the process must be in some state after it leaves states i, these transition probabilities satisfy

$$\sum_{j=1}^{N} P_{ij} = 1, \qquad i = 1, \ldots, N$$

A Markov chain is said to be irreducible if for each pair of states i and j there is a positive probability, starting in state i, that the process will ever enter state j. For an irreducible Markov chain, let π_j denote the long-run proportion of time that the process is in state j. (It can be shown that π_j exists and is constant, with probability 1, independent of the initial state.) The quantities π_j, $j = 1, \ldots, N$, can be shown to be the unique solution of the following set of linear equations:

$$\pi_j = \sum_{i=1}^{N} \pi_i P_{ij}, \qquad j = 1, \ldots, N \tag{10.4}$$

$$\sum_{j=1}^{N} \pi_j = 1 \tag{10.5}$$

REMARK

The above set of equations have a heuristic interpretation. Since π_i is the proportion of time that the Markov chain is in state i and since each transition out of state i is into state j with probability P_{ij}, it follows that $\pi_i P_{ij}$ is the proportion of time that the Markov chain has just entered state j from state i. Hence Equation (10.4) states the intuitively clear fact that the proportion of time that the Markov chain has just entered state j is equal to the sum over all states i, of the proportion of time it has just entered state j from state i. Equation (10.5) says, of course, that by summing over all j the proportion of time that the chain is in state j must equal 1. ■

The following example provides the theoretical basis of the Markov chain simulation method.

Example 10b Consider a set of N elements, called nodes, and suppose that certain pairs of nodes are connected by a line. If nodes i and j are connected by a line, then we say that the pair (i, j) constitutes an arc. The set of nodes along with the set of arcs is called a graph. In Figure 10.3 the graph consists of the nodes 1, 2, 3, 4, 5 along with the arcs (1, 2), (1, 3), (1, 5), (2, 4), (3, 4). It should be noted that there is no ordering of the arcs; that is, the arc (1, 2) is the same as the arc (2, 1).

Consider now a Markov chain that moves along the nodes of a graph in the following manner: whenever it is at node i the next node is equally likely to be any of the neighbors of i, where we say that j is a neighbor of i if (i, j) is an arc of the graph. For example, in Figure 10.3 the transition probabilities of the resulting Markov chain are

$$P_{12} = P_{13} = P_{15} = \tfrac{1}{3}, \qquad P_{21} = P_{24} = \tfrac{1}{2}$$
$$P_{34} = P_{31} = \tfrac{1}{2}, \qquad P_{42} = P_{43} = \tfrac{1}{2}, \qquad P_{51} = 1$$

Suppose now that this Markov chain is irreducible [which is equivalent to assuming that for any pair of nodes i and j there is a path that goes from i to j—that is, a sequence of arcs of the form (i, i_1), (i_1, i_2), . . . , (i_{k-1}, i_k), (i_k, j)]. Let A denote the set of arcs.

If we let d_i, called the degree of node i, denote the number of arcs leading out of node i, then the transition probabilities of the Markov chain can be expressed as follows:

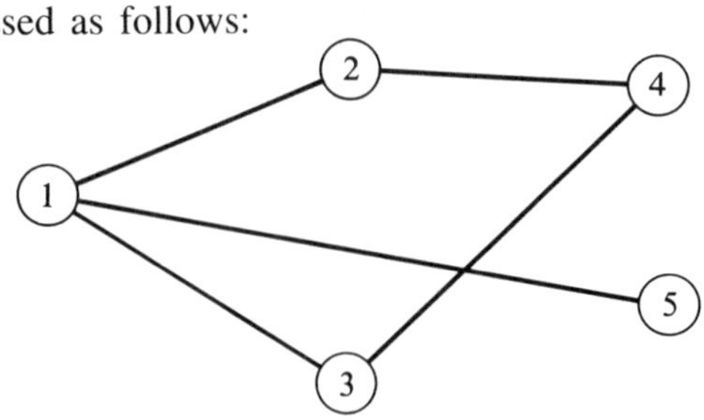

Figure 10.3 A graph.

$$P_{ij} = \begin{cases} \dfrac{1}{d_i} & \text{if } (i, j) \text{ is an arc} \\ 0 & \text{otherwise} \end{cases}$$

The long-run proportions of time that the process is in each of the states—that is, the π_i—are thus the unique solutions to the set of equations

$$\pi_j = \sum_{i:(i,j)\in A} \frac{\pi_i}{d_i}, \qquad j = 1, \ldots, N$$

$$\sum_j \pi_j = 1$$

If we try a solution of the form $\pi_j = cd_j$, the above equations reduce to

$$cd_j = \sum_{i:(i,j)\in A} c = cd(j)$$

$$\sum_j cd_j = 1$$

That is,

$$\pi_j = \frac{d_j}{\sum_j d_j}$$

satisfies the equations. Hence, by uniqueness, it follows that these values represent the long-run proportions of time that the process is at each of the nodes. For example, in the graph given by Figure 10.1 since $d_1 = 3$, $d_2 = d_3 = d_4 = 2$, and $d_5 = 1$, we see that $\pi_1 = \frac{3}{10}$, $\pi_2 = \pi_3 = \pi_4 = \frac{1}{5}$, and $\pi_5 = \frac{1}{10}$. ■

Suppose now that we are interested in determining the average value of some function defined over a large and complicated "combinatorial set" C. For example, C might be the set of all permutations $(x_1, \ldots, x_n)$ of the numbers $(1, 2, \ldots, n)$ for which $\sum_{j=1}^n jx_j \leq k$ for a given constant k; or C might be the set of all subgraphs of a given graph having the property that for any pair of nodes i and j there is a unique path in the subgraph from i to j (such a subgraph is called a tree).

Suppose that for a given function h over the elements of the set C we are interested in determining θ, the average value of $h(X)$, where X is equally likely to be any of the elements of the set. That is,

$$\theta = \frac{\sum_{x \in C} h(x)}{N}$$

where N is the number of elements of C. Because N may be very large, and indeed its value is often hard to determine, how can we estimate θ?

The standard simulation approach would be to generate a random element X from C that is equally likely to take on any of the N values and then compute $h(X)$. This operation is repeated k times and the average of the values obtained is then taken as our estimate of θ, where k is determined by our ongoing estimate of the variance of the estimator. However, in the above situation it is usually not at all apparent how one can generate a random element from C and thus this approach may not be available.

The Markov chain simulation approach for estimating θ operates by first defining a concept of neighboring elements in C. It then considers the resulting graph obtained from the set of nodes C when an arc is said to exist between each pair of neighboring nodes. For example, if C is the set of permutations of (1, 2, 3, 4, 5), we can define two permutations to be neighbors if one results from an interchange of two of the positions of the other. That is, (1, 2, 3, 4, 5) and (1, 4, 3, 2, 5) are neighbors, whereas (1, 2, 3, 4, 5) and (1, 4, 2, 3, 5) are not. A similar definition of neighbors can be used when C is the set of permutations $(x_1, \ldots, x_n)$ satisfying the condition that $\sum_{j=1}^{n} jx_j \leq k$. If C is a set of trees, we can say that two trees are neighbors if all but one of the arcs of the first tree are arcs of the second tree.

Suppose that, starting with an arbitrary element in C, we now simulate the Markov chain which is at each step equally likely to move from the present node to either of its neighbors (as in Example 10b). If X_n denotes the nth state visited, then, letting $\pi(x)$ denote the long-run proportion of time the state is x, it follows that, for any function f on the elements of C, the long-run average value of $f(X_n)$ is $\sum_{x \in C} \pi(x)f(x)$. That is,

$$\sum_{n=1}^{k} \frac{f(X_n)}{k} \to \sum_{x \in C} \pi(x)f(x) \qquad \text{as } k \to \infty$$

Hence, since in Example 10b we obtained that $\pi(x) = d(x)/D$, where $d(x)$ is the number of neighbors of x, and $D = \sum_{x \in C} d(x)$, we see that

$$\frac{\sum_{n=1}^{k} f(X_n)}{k} \to \sum_{x \in C} f(x)\frac{d(x)}{D} \tag{10.6}$$

If we now take f to be the function specified by

$$f(x) = \frac{h(x)}{d(x)}$$

we obtain from Equation (10.6) that

$$\frac{1}{k}\sum_{n=1}^{k}\frac{h(X_n)}{d(X_n)} \to \sum_{x\in C}\frac{h(x)}{D} \tag{10.7}$$

Similarly, if we let

$$f(x) = \frac{1}{d(x)}$$

we obtain from (10.6) that

$$\frac{1}{k}\sum_{n=1}^{k}\frac{1}{d(X_n)} \to \sum_{x\in C}\frac{1}{D} = \frac{N}{D} \tag{10.8}$$

Hence, from Equations (10.7) and (10.8) we see that

$$\frac{\sum_{n=1}^{k}\dfrac{h(X_n)}{d(X_n)}}{\sum_{n=1}^{k}\dfrac{1}{d(X_n)}} \to \frac{\sum_{x\in C} h(x)}{N} = \theta$$

The above represents the Markov chain simulation method. To estimate θ it first simulates the Markov chain determined from the graph of ''neighbors'' and then estimates θ by dividing the average value of $h(X_n)/d(X_n)$ obtained by the average value of $1/d(X_n)$.

10.4. Estimating Probabilities and Expected First Passage Times by Using Random Hazards

Consider a process that moves from state to state in the following manner. If the present state is x, then, independent of the sequence of previous states, the next state is chosen according to the distribution function F_x. Such a process is called a Markov process (if the state space is discrete, it is called a Markov chain).

For a given initial state—call it 0—and a fixed set of states A, suppose we are interested in estimating the expected number of transitions needed for the process to enter the set A. That is, if X_n is the state at time n, then we are interested in $E[N]$, where

$$N = \text{Min}\{n\colon\ X_n \in A\}$$

In many cases of interest A is a ''rare set of states'' in the sense the process has a tendency to be drawn to the initial state 0 with visits to A occurring infrequently. Hence, in such cases, $E[N]$ is relatively large. In addition, since the process has this tendency to revert back to 0, it follows, by the memoryless property, that N will roughly have an exponential distribution (see Example 8a of Chapter 8 for a more detailed discussion of this latter point). But if such is the case, we have that $\text{Var}(N) \approx (E[N])^2$, implying that $\sqrt{\text{Var}(N)}$, the standard deviation of N, is also large; and so the raw simulation approach to estimate $E[N]$ requires a huge number of runs.

A powerful variance reduction technique makes use of the so-called observed hazards, also called the predictable projections, as control variates. It works as follows. Consider a simulation run that results in the sequence of states $X_0, \ldots, X_N$ and define the observed hazards, λ_n, $n \geq 0$, by

$$\lambda_n = P\{N = n | X_0, \ldots, X_{n-1}\}$$

In words, if A has not yet been entered before time n, then λ_n is the probability, given the state at time $n - 1$, that the next state will be in the set A. Now let Y denote the sum of these hazards—that is,

$$Y = \sum_{n=1}^{N} \lambda_n$$

Proposition:

$$E\left[\sum_{n=1}^{N} \lambda_n\right] = P\{N < \infty\}.$$

Proof. Let

$$I_n = \begin{cases} 1 & \text{if } N = n \\ 0 & \text{otherwise} \end{cases}$$

and note that

$$E[I_n | X_0, \ldots, X_{n-1}] = P\{N = n | X_0, \ldots, X_{n-1}\} = \lambda_n$$

Now

$$\sum_{n=1}^{\infty} I_n = \begin{cases} 1 & \text{if } N < \infty \\ 0 & \text{otherwise} \end{cases}$$

and so, by taking expectations,

$$
\begin{aligned}
P\{N < \infty\} &= E\left[\sum_{n=1}^{\infty} I_n\right] \\
&= \sum_{n=1}^{\infty} E[I_n] \\
&= \sum_{n=1}^{\infty} E[E[I_n|X_0, \ldots, X_{n-1}]] \\
&= E\left[\sum_{n=1}^{\infty} E\left[I_n|X_0, \ldots, X_{n-1}\right]\right] \\
&= E\left[\sum_{n=1}^{\infty} \lambda_n\right] \\
&= E\left[\sum_{n=1}^{N} \lambda_n\right] \qquad \square
\end{aligned}
$$

Suppose now that the situation is such that A is eventually entered with probability 1, and so the expected sum of the random hazards has mean 1. Since runs having large values of N often result from many missed opportunities to enter A, it follows that Y is a natural candidate for a control variable since it gives a way of gauging whether a given value of N is much larger or smaller than normal. Therefore we should simulate until the process enters A and then use the output estimator

$$N + c(Y - 1)$$

The best value of c, given by

$$c = -\frac{\text{Cov}(N, Y)}{\text{Var}(Y)}$$

can be approximated by a preliminary short simulation.

Example 10c: Cumulative Sum Control Chart for Exponential Random Variables. Let $X_1, X_2, \ldots$ denote a sequence of independent exponential random variables each with mean 1. Let $S_0 = 0$ and define

$$S_n = \text{Max}\{0, S_{n-1} + X_n - 2\}, \qquad n \geq 1$$

We are interested in estimating $E[N]$, where

$$N = \text{Min}\{n\colon\ S_n > 4\}$$

The above sequence S_n, $n \geq 1$, is called a cumulative sum control chart. The idea is that the quantities X_n represent the successive values of items produced in a given process. When the process is in control, these values are assumed to have an exponential distribution with mean 1; whereas when the process goes out of control these values tend to become larger. The quantities S_n are cumulative sums of the successive quantities $X_n - 2$, which are not allowed to become negative—that is, the value of the cumulative sum is reset to 0 whenever it would fall negative. Since the values $X_n - 2$ have a negative mean when the process is in control, the cumulative sum usually has a small value in this case and so a large value is an indicator that the process may have gone out of control. However, since large values eventually occur even when the process remains in control, we are interested in $E[N]$, the expected time until the process is mistakenly declared out of control when in fact it remains in control throughout.

Since the values S_n tend to be near 0, excursion to a value as large as 4 tends to be a rare event whose distribution is approximately exponential. Thus the raw simulation estimator has a large standard deviation. The random hazard is given by

$$\begin{aligned}\lambda_n &= P\{S_{n-1} + X_n - 2 > 4 | S_{n-1}\}\\ &= P\{X_n > 6 - S_{n-1} | S_{n-1}\}\\ &= \exp\{S_{n-1} - 6\}\end{aligned}$$

Hence the use of the controlled estimator

$$N + c[Y - 1]$$

where

$$Y = \sum_{n=1}^{N} \exp\{S_{n-1} - 6\}$$

is recommended. Indeed, a small simulation study, in which part of the simulation was used to estimate the best c—namely,

$$c^* = -\frac{\text{Cov}(N, Y)}{\text{Var}(Y)}$$

revealed to the author that the variance of the controlled estimator is smaller than that of the raw estimator by a factor of roughly $\frac{1}{30}$. In addition, it was seen that $E[N] \approx 243$, and that the improvement was rather robust for any value of c used which was near $-E[N] = -243$.

Example 10d: Expected Length of a Busy Period in a Queueing System. Consider a single server queue in which the times between arriving customers are independent random variables having a common distribution F. (In other words, the arrival process is a renewal process: see Section 8.3 of Chapter 8.) There is a single server that takes a random time to serve a customer. An arrival finding the server idle immediately enters service, whereas one finding the server busy joins the waiting queue. Upon a service completion, one of the waiting customers, if there are any, enters service. Each service is assumed to take a random time, independent of the past, having distribution G.

Such a system as described above alternates between times when the server is busy and times when he is idle. Suppose we are interested in determining $E[N]$, the expected number of customers served in an arbitrary busy period.

Let W_n denote the total amount of work that remains in the system—that is, the sum of the service times of those customers waiting plus the remaining service time of the customer in service—immediately after the arrival of the nth customer. Also, let X_n denote the number of customers in the system found by the nth arrival, so that $X_1 = 0$ and

$$N = \text{Min}\{n > 1: X_n = 0\} - 1$$

Now consider the process whose state at time n is the pair W_n, X_n and define the random hazards by

$$\lambda_n = P\{N = n | W_1, X_1, \ldots, W_n, X_n\}$$

Now given that the workload immediately after the nth arrival is w, it follows that the next arrival will find the system empty—implying that $N = n$—if the next interarrival time is greater than w. Hence, for $n \leq N$,

$$\lambda_n = 1 - F(W_n) = \overline{F}(W_n)$$

Thus we recommend using the estimator

$$N + c\left[\sum_{n=1}^{N} \overline{F}(W_n) - 1\right]$$

where the value of c can be determined from the simulation.

If the queueing system contained more than one server, we could still proceed as above except we would now define W_n as the time until the system becomes empty of all customers presently in the system at the moment immediately after the nth customer arrives. The estimator would then be as given above.

If we wanted to estimate $E[T]$, the mean time of a busy period, we could either estimate $E[N]$ and use the identity

$$E[T] = E[N]\mu_F$$

where μ_F is the mean time between arrivals, or we could directly use the estimator T—again with the above as a control. That is, we could use

$$T + c\left[\sum_{n=1}^{N} \overline{F}(W_n) - 1\right]$$

with c now equal to the negative of the covariance between T and the control divided by the variance of the control. Whereas it is not immediately clear which of these approaches is best (either estimating $E[N]$ and then using $E[N]\mu_F$ or directly estimating $E[T]$), the author expects that the former method is usually better. ■

Since

$$E\left[\sum_{n=1}^{N} \lambda_n\right] = P\{N < \infty\}$$

we can also use the sum of the hazards as a direct estimator of $p \equiv P\{N < \infty\}$ in cases where this probability is less than 1. For example, let us reconsider Example 10c. Another approach to using simulation to estimate $E[N]$ is to define a cycle to occur either when S_n exceeds 4 or when it returns to 0, and let C denote the time of a cycle. Then it is easy to see that

$$E[N] = \frac{E[C]}{p}$$

where p is the probability that a value greater than 4 occurs before the random walk goes negative. Since p tends to be small, it would seem that we could efficiently estimate it by using the sum of the hazards of N during a cycle as the estimator. That is, we could use the estimator

$$\sum_{n=1}^{C} \lambda_n = \sum_{n=1}^{C} \exp\{S_{n-1} - 6\}$$

In addition, because $E[C]$ is probably relatively small, it can either be estimated by the raw simulation estimator or we could use the sum of the hazards for C as a control. That is, the latter case suggests that $E[C]$ be estimated by an estimator of the form

$$C + c\left[\sum_{n=1}^{C} \left(P\left\{X_n > 6 - S_{n-1}\right\} + P\left\{X_n < 2 - S_{n-1}\right\}\right) - 1\right]$$

For another example illustrating the direct use of the sum of the hazards as an estimator of a probability, suppose we wanted to estimate the reliability function of a system of n components (see Example 8b of Chapter 8). That is, let $S_1, \ldots, S_n$ be independent Bernoulli random variables with

$$p_i = P\{S_i = 1\}$$

and for a given binary increasing function $\phi(s_1, \ldots, s_n)$ suppose we are interested in computing

$$p \equiv E[\phi(S_1, \ldots, S_n)] = P\{\phi(S_1, \ldots, S_n) = 1\}$$

In many cases of interest the system is "highly reliable" and so p is quite large, or, equivalently, the failure probability $q = 1 - p$ is quite small; and as it would be important in such cases to distinguish such values of p as .9900 and .9999, we would need a large amount of simulation when using the raw estimator.

We can apply the random hazard concept in the above reliability application by generating the S_i sequentially, $i = 1, \ldots, n$, and considering the resulting Markov chain $\{X_k\}$, where

$$X_k = (S_1, \ldots, S_k)$$

and where N is defined to be the first passage to a state $(s_1, \ldots, s_k)$ for which $\phi(s_1, \ldots, s_k, 1, 1, \ldots, 1) = 0$. That is, N is the first time we are certain that the system is failed under the generated state. Hence, for $N > k - 1$, the random hazards are given by

$$\begin{aligned}\lambda_k &= P\{\phi(S_1, \ldots, S_k, 1, 1, \ldots, 1) = 0 | S_1, \ldots, S_{k-1}\} \\ &= \begin{cases} q_k & \text{if } S_k \text{ is critical given } S_1, \ldots, S_{k-1} \\ 0 & \text{otherwise} \end{cases}\end{aligned}$$

where $q_i = 1 - p_i$, and where by critical we mean that the system necessarily fails if component k is failed.

Of course, the above is not necessarily the best way to define the Markov chain. For example, it might be better not to consider the components one at a time but to consider them in groups so that there is always a possibility that the next state of the Markov chain will enter the failed set A. That is, suppose that the system will fail if at least r of the n components are failed. It might now be best to consider the Markov chain for which from any state $S_1, \ldots, S_k$ which has i failed components the next state consists of $S_1, \ldots, S_k, S_{k+1}, \ldots, S_{k+r-i}$. In doing it this way there is always a positive probability that the next state will be in the failed set. In addition, it would seem that we should start, in this example, by reordering the components in decreasing order of their failure probabilities q_i. The estimate we would use from a simulation run is the sum of the hazards.

Example 10e Consider a 3-of-5 system that fails if at least 3 of the 5 components are failed, and suppose that each component fails with probability $q = 1 - p$. If we now generate the states of these five components and use the Markov chain, which looks at them in groups so that there would be exactly three failures if all members of the group under consideration are failed, then the total hazard is

$$\begin{array}{lll} q^3 & \text{with probability} & p^3 + q^3 \\ q^3 + q^2 & \text{with probability} & 3p^2q \\ q^3 + q & \text{with probability} & 3pq^3 \\ q^3 + 2q & \text{with probability} & 3p^2q^2 \end{array}$$

The first case follows since we initially look at components 1, 2, 3 and so the first hazard value is q^3; and if these three components are either all failed (implying that the system is failed) or are all working (implying that the system is working) then there are no additional hazards. If exactly one of the first three components is failed—and this occurs with probability $3p^2q$—then we then consider the final two components and so the second (and final) hazard in this case is q^2. The third possibility occurs if the first three components contain two failures and the next component is also a failure (and the fourth when it is a success).

If we now set $q = 0.1$, we see that

$$E[\text{total hazard}] = .00856, \qquad \text{Var(total hazard)} = .000966$$

Since

$$E[\text{total hazard}] = P\{\text{system is failed}\}$$

we see that the above variance is much smaller than the variance of the Bernoulli raw simulation estimator I, which has

$$\text{Var}(I) = .00856(1 - .00856) = .008487$$

Indeed,

$$\frac{\text{Var}(I)}{\text{Var(total hazard)}} = 8.784101$$

Table 10.1 gives the ratios of $\text{Var}(I)$ to the variance of the total hazard for a variety of values of q.

Table 10.1. 3-of-5 System

q	.001	.01	.05	.1	.2	.3	.4
$\dfrac{\text{Var}(I)}{\text{Var(total hazard)}}$	833.34	83.39	16.92	8.78	4.88	3.61	2.86

REMARK

If we had considered the components one at a time then, for $q = 0.1$, the variance of the total hazard would have been $\mathrm{Var}(\Sigma_n \lambda_n) = .001611$, which is better than that of the raw estimator but not as good as the total hazard estimator used above. ■

Problems

1. Set up the alias method for generating a binomial with parameters (5, .4).
2. Explain how we can number the $\mathbf{Q}^{(k)}$ in the alias method so that k is one of the two points to which $\mathbf{Q}^{(k)}$ gives weight.
3. Complete the details of Example 10a.
4. Write a program to generate the points of a two-dimensional Poisson process within a circle of radius R, and run the program for $\lambda = 1$ and $R = 5$. Plot the points obtained.
5. Let C denote all permutations of $1, 2, \ldots, n$ and say that any two permutations that differ by a single interchange are neighbors.
   (a) How many neighbors does each permutation have?
   (b) Explain, for a given permutation, how you could generate a new permutation that is equally likely to be any of its neighbors.
6. Consider all permutations $(P_1, \ldots, P_{10})$ of $(1, 2, \ldots, 10)$ such that $\Sigma_j jP_j \leq 300$. Use simulation to estimate the average value of $\Sigma_j j^2P_j$.
7. Let $Z_1, Z_2, \ldots$ denote a sequence of independent unit normals. Let

$$M_n = \frac{(Z_n + Z_{n-1} + Z_{n-2} + Z_{n-3})}{4}, \qquad n \geq 4$$

   and define N by

$$N = \mathrm{Min}\{n: |M_n| > \tfrac{3}{2}\}$$

   We want to use simulation to find $E[N]$. Determine the variance of the raw simulation estimator and then of the one that uses the sum of the random hazards as a control. (The above is called a moving-average control chart and is used to determine when the distribution has shifted from the unit normal.)
8. In Example 10e show that if the components are considered one at a time then $\mathrm{Var}(E_n\lambda_n) = .001611$ when $q = 0.1$.

Appendix of Programs

PROGRAM 4-1

```
10 PRINT ''THIS PROGRAM WILL SIMULATE N IID POISSON RANDOM VARI-
     ABLES''
20 PRINT ''ENTER THE MEAN''
30 INPUT L
40 PRINT ''ENTER N''
50 INPUT N
60 RANDOMIZE
70 I=INT(L)
80 B=1/L
90 FOR K=1 TO I
100 S=S+LOG(K)
110 NEXT
120 S=-S-L+I*LOG(L)
130 PP=EXP(S)
140 F=1
150 FOR K=1 TO I
160 F=F*(I+1-K)*B
170 CUM=CUM+F
```

```
180 NEXT
190 CUM=(CUM+1)*PP
200 FOR J=1 TO N
210 I=INT(L)
220 A=CUM
230 P=PP
240 U=RND
250 IF U<A GOTO 310
260 I=I+1
270 P=L*P/I
280 A=A+P
290 IF U<A THEN GOTO 360
300 GOTO 260
310 A=A-P
320 IF U>A GOTO 360
330 P=I*P*B
340 I=I-1
350 GOTO 310
360 PRINT I
370 NEXT
380 END
```

PROGRAM 9-0

```
10 PRINT ''THIS PROGRAM COMPUTES THE PROBABILITY THAT A UNIT NORMAL
     RANDOM VARIABLE IS LESS THAN X''
20 PRINT ''ENTER THE DESIRED VALUE OF X''
30 INPUT X
40 U-ABS(X)
50 IF U>4 GOTO 180
60 Y=U^2
70 I=U
80 FOR J=1 TO 40
90 U=-U*Y*(2*J-1)/(2*J*(2*J+1))
100 I=I+U
110 NEXT
120 I=I/SQR(2*3.14159)
130 IF X<0 GOTO 160
140 PRINT ''THE PROBABILITY IS''.5+I
150 GOTO 220
160 PRINT ''THE PROBABILITY IS''.5-I
170 GOTO 220
180 IF X<0 GOTO 210
190 PRINT ''THE PROBABILITY IS GREATER THAN'' 1-10^-4
200 GOTO 220
210 PRINT ''THE PROBABILITY IS LESS THAN'' 10^-4
220 END
```

PROGRAM 9-1

```
10 PRINT ''THIS PROGRAM COMPUTES THE PROBABILITY THAT A CHI-SQUARE
     RANDOM VARIABLE WITH N DEGREES OF FREEDOM IS LESS THAN X''
20 PRINT ''ENTER THE DEGREE OF FREEDOM PARAMETER''
30 INPUT N
40 S=(N-1)/2
50 PRINT ''ENTER THE DESIRED VALUE OF X''
```

```
 60 INPUT X
 70 M=X/2
 80 D=X/2-N/2+1/3
 90 D=D-.04/N
100 IF N=1 GOTO 160
110 IF S=M GOTO 180
120 H=S/M
130 X=(1-H*H+2*H*LOG(H))/(1-H)^2
140 X=D*SQR((1+X)/M)
150 GOTO 190
160 X=D*SQR(2/M)
170 GOTO 190
180 X=D/SQR(M)
190 U=ABS(X)
200 IF U>4 GOTO 330
210 Y=U^2
220 I=U
230 FOR J=1 TO 40
240 U=-U*Y*(2*J-1)/(2*J*(2*J+1))
250 I=I+U
260 NEXT
270 I=I/SQR(2*3.14159)
280 IF X<0 GOTO 310
290 PRINT "THE PROBABILITY IS".5+I
300 GOTO 370
310 PRINT "THE PROBABILITY IS".5-I
320 GOTO 370
330 IF X<0 GOTO 360
340 PRINT "THE PROBABILITY IS GREATER THAN" 1-10^-4
350 GOTO 370
360 PRINT "THE PROBABILITY IS LESS THAN" 10^-4
370 END
```

PROGRAM 9-2

```
 10 PRINT "THIS PROGRAM USES SIMULATION TO APPROXIMATE THE p-VALUE
       IN THE GOODNESS OF FIT TEST"
 20 RANDOMIZE
 30 PRINT "ENTER THE NUMBER OF POSSIBLE VALUES"
 40 INPUT N
 50 DIM P(N)
 60 DIM Q(N)
 66 Q(0)=0
 70 PRINT "ENTER THE PROBABILITIES ONE AT A TIME"
 80 FOR I=1 TO N
 90 INPUT P(I)
100 Q(I)=Q(I-1)+P(I)
110 NEXT I
120 PRINT "ENTER THE SAMPLE SIZE"
130 INPUT D
140 DIM B(N)
150 DIM C(N)
160 FOR J=1 TO N
170 B(J) = D*P(J)
180 C(J)=1/B(J)
190 NEXT
200 DIM X(N)
```

```
210 PRINT ''ENTER THE DESIRED NUMBER OF SIMULATION RUNS''
220 INPUT R
230 PRINT ''ENTER THE VALUE OF THE TEST STATISTIC''
240 INPUT W
250 FOR K=1 TO R
260 FOR L=1 TO N
270 X(L)=0
280 NEXT L
290 FOR J=1 TO D
300 U=RND
310 I=1
320 IF U<Q(I) THEN GOTO 350
330 I=I+1
340 GOTO 320
350 X(I)=X(I)+1
360 NEXT J
370 S=0
380 FOR L=1 TO N
390 S=S+(X(L)-B(L))^2*C(L)
400 NEXT L
410 IF S>=W THEN C=C+1
420 NEXT K
430 PRINT ''THE ESTIMATE OF THE p-VALUE IS'' C/R
440 END
```

PROGRAM 9-3

```
 10 PRINT ''THIS PROGRAM USES SIMULATION TO APPROXIMATE THE p-VALUE
       OF THE KOLMOGOROV-SMIRNOV TEST''
 20 RANDOMIZE
 30 PRINT ''ENTER THE VALUE OF THE TEST QUANTITY''
 40 INPUT D
 50 PRINT ''ENTER THE SAMPLE SIZE''
 60 INPUT N
 70 A=1/N
 80 PRINT ''ENTER THE DESIRED NUMBER OF SIMULATION RUNS''
 90 INPUT R
100 DIM T(N)
110 FOR I=1 TO R
120 S=0
130 T(0) = 0
140 FOR J=1 TO N
150 U=RND
160 Y = -LOG(U)
170 T(J) = T(J-1)+Y
180 S=S+Y
190 NEXT
200 S = S-LOG(RND)
210 C=1/S
220 J=1
230 X = T(J)*C
240 IF J*A - X > D THEN GOTO 290
250 IF X - (J-1)*A > D THEN GOTO 290
260 IF J=N GOTO 300
270 J=J+1
280 GOTO 230
290 CC=CC+1
```

```
300 NEXT I
310 PRINT ''THE APPROXIMATE p-VALUE IS'' CC/R
320 END
```

PROGRAM 9-4

```
 10 PRINT ''THIS PROGRAM COMPUTES THE p-VALUE FOR THE TWO-SAMPLE RANK
      SUM TEST''
 20 PRINT ''THIS PROGRAM WILL RUN FASTEST IF YOU DESIGNATE AS THE FIRST
      SAMPLE THE SAMPLE HAVING THE SMALLER SUM OF RANKS''
 30 PRINT ''ENTER THE SIZE OF THE FIRST SAMPLE''
 40 INPUT N
 50 PRINT ''ENTER THE SIZE OF THE SECOND SAMPLE''
 60 INPUT M
 70 PRINT ''ENTER THE SUM OF THE RANKS OF THE FIRST SAMPLE''
 80 INPUT T
 90 DIM P(N,M,T+1)
100 FOR I=1 TO N
110 FOR K=I*(I+1)/2 TO T
120 P(I,0,K)=1
130 NEXT
140 NEXT
150 FOR K=1 TO T+1
160 FOR J=1 TO M
170 P(0,J,K-1)=1
180 NEXT
190 NEXT
200 FOR I=1 TO N
210 FOR J=1 TO M
220 FOR K=1 TO T
230 IF K<(I+J) THEN P(I,J,K)=(J/(I+J))*P(I,J-1,K) ELSE
      P(I,J,K)=(I/(I+J))*P(I-1,J,K-I-J)+(J/(I+J))*P(I,J-1,K)
240 NEXT
250 NEXT
260 NEXT
270 IF P(N,M,T) < 1-P(N,M,T-1) THEN V=P(N,M,T) ELSE
      V=1-P (N,M,T-1)
280 PRINT ''THE p-VALUE IS'' 2*V
290 END
```

PROGRAM 9-5

```
 10 PRINT ''THIS PROGRAM APPROXIMATES THE p-VALUE IN THE TWO-SAMPLE
      RANK SUM TEST BY A SIMULATION STUDY''
 20 RANDOMIZE
 30 PRINT ''ENTER THE SIZE OF THE FIRST SAMPLE''
 40 INPUT N(1)
 50 PRINT ''ENTER THE SIZE OF THE SECOND SAMPLE''
 60 INPUT N(2)
 70 PRINT ''ENTER THE SUM OF THE RANKS OF THE FIRST SAMPLE''
 80 INPUT T
 90 PRINT ''ENTER THE DESIRED NUMBER OF SIMULATION RUNS''
100 INPUT M
110 N=N(1)+N(2)
120 DIM X(N)
130 NUM=NUM+1
```

```
140 S=0
150 FOR I=1 TO N
160 X(I)=I
170 NEXT I
180 FOR I=1 TO N(1)
190 R=INT((N+1-I)*RND)+1
200 S=S+X(R)
210 X(R)=X(N+1-I)
220 NEXT I
230 IF S<=T THEN C(1)=C(1)+1
240 IF S>=T THEN C(2)=C(2)+1
250 IF NUM < M GOTO 130
260 IF C(1)>C(2) THEN C(1)=C(2)
270 PRINT "THE APPROXIMATE p-VALUE IS" 2*C(1)/M
280 END
```

Index